# Position Sensing

# Position Sensing
## Angle and Distance Measurement
## for Engineers

**Dr Ing Hans Walcher**

Translated by
David Kerr, BA
and
M. J. Shields, FIInfSc, MITI

BUTTERWORTH
HEINEMANN

Butterworth-Heinemann Ltd
Linacre House, Jordan Hill, Oxford OX2 8DP

$\mathcal{R}$  A member of the Reed Elsevier group

OXFORD   LONDON   BOSTON
MUNICH   NEW DELHI   SINGAPORE   SYDNEY
TOKYO   TORONTO   WELLINGTON

First published by VDI Verlag 1985
First published in Great Britain by Butterworth-Heinemann Ltd 1994

**British Library Cataloguing in Publication Data**
Walcher, Hans
    Position Sensing: Angle and Distance
    Measurement for Engineers. – 2 Rev.ed
    I. Title II. Kerr, David III. Shields,
    Michael J.
    620.0044

ISBN 0 7506 1157 X

**Library of Congress Cataloguing in Publication Data**
Walcher, Hans.
    [Digitale Lagemesstechnic.   English]
    Position sensing: angle and distance measurement for engineers/
    Hans Walcher; translated by David Kerr and M.J. Shields.
        p.    cm.
    Includes bibliographical references and index.
    ISBN 0 7506 1157 X
    1. Electric measurements.   2. Physical measurements.   3. Digital
    electronics.   4. Angle–Measurement.   5. Distances–Measurement.
    I. Title.
    TK277.W3213                                            93-41284
    681'.2–dc20                                                 CIP

Typeset by Create Publishing Ltd, Bath, Avon.
Printed in Great Britain by Redwood Books, Trowbridge, Wiltshire

# Contents

# *Preface*

The electrical measurement of angles and distances is an interesting interdisciplinary field as widely differing physical principles can be used to produce the measurement base and its reading and to evaluate and process the signals obtained.

This volume is intended both for the mechanical engineer, who has to include measurement processes as an integral part of his design, and the measurement and control technician who, in widely differing engineering tasks, may be confronted with the problem of position sensing, often in relation to positioning control. In many cases their interest will be limited to the electrical interface, the method and the security of data transfer and storage.

When the first edition of this book was published in 1974 under the title *Digitale Lagemesstechnik*, the electrical measurement of angles and distances in engineering was severely restricted to lathes and milling machines with numerical control systems. The users of these numerically-controlled machines were mainly large companies, for example in the aircraft and motor industries, who were able to purchase and operate this costly and by no means always reliable technology. The measurement systems used, which in most cases consisted of analogue transducers backed by analogue-to-digital converters, were of interest only to a small band of specialists.

Since then, the digital display of position data and the digital control that comes with it have been commonplace even in small and medium-sized businesses. A decisive part in this process was played by the microprocessor and the development of increasingly efficient and, more importantly, cheaper controllers. The widespread availability of lower-priced controllers has meant that measurement processes have also become available to a widening range of manufacturers and users. Today there is hardly a branch of engineering in which automatic measurement and control does not take place. Examples of this are machines for woodworking, metalworking and processing plastics, printing machines, packing machines, food processing machines, etc. In addition, high-precision automatic position-sensing systems are being used to sense and control the position of radio telescopes, astronomical telescopes and radar antennae, to control integral switching in production systems, and in automatic measuring and testing machines. In all branches of technology where controlled movement takes place, position sensing is at the forefront. Further examples are the navigation of ships, vehicles and aircraft, and the control of conveyor and storage systems.

In the second edition which was completely revised and greatly enlarged in 1985 under the title *Winkel- und Wegmessung im Maschinenbau*, not only was consideration given to the technical evolution which had taken place in the meantime, but analogue electrical systems such as potentiometers, inductive and capacitive transducers and mechanical measurement processes were also included. In addition, newer mechanical position measuring systems have clearly justified their existence with a good cost/benefit ratio and are widely used.

In the years between 1985 and 1992 the requirements for resolution and accuracy in distance measurement processes have continued to become more stringent. Optically scanned scales with phase gratings are described as a typical example. Moreover, there has been a renaissance in resolver technology, the advantages of which have been recognised when used at higher operating temperatures. A subject increasingly under discussion in technical circles is what is known as intelligent sensors, to which a chapter has been devoted.

All measurement processes are usually described in terms of the basic principles applied. Switching details are only described where they are necessary to acquire an understanding of the function.

The German edition of my book, which to my knowledge is the only comprehensive treatment of this interesting and important area of measurement engineering, was published by VDI Verlag. I am delighted that the publishers Butterworth-Heinemann are now making the book accessible to a much wider readership by bringing out an English-language addition.

At this point I would like to thank all the companies who have assisted me in the preparation of this volume by providing photographic material. I would especially like to mention Erwin Halstrup

Multur GmbH without whom this new edition would not have been published. I also thank my secretary Frau Rose Riedl in equal measure for her assistance. My special thanks are due to my wife for her patience with this time-consuming hobby.

Hans Walcher
Kirchzarten
February 1992

# 1

## Introduction

The measurement of angle and distance is a basic requirement in all instruments, machines and installations in which position has to be monitored or adjusted. There are still many machines which use simple analogue measuring systems. In the past, these usually consisted of a scale and a mark or index which moved relative to it; the scale could be in the form of a ruler, a disc or a drum. As it is only a matter of the relative movement between the scale and mark, the scale can be fixed and the index movable or the scale movable and the index fixed. Recently, however, analogue mechanical indicators have increasingly been replaced by digital mechanical counters which give more reliable readout and more accurate measurement. Because of the mechanics involved, these positional indications are limited to medium and low-speed applications, i.e. mostly on manually-operated machine arbors, which in the course of automation are increasingly being replaced by motorized and numerically-controlled arbors. Spindles for adjusting limit stops, supports, tools, etc., which are set up before actual production begins (setting procedure) are an exception. Digital mechanical position indicators are very often used for cost reasons.

High resolutions and high accuracies can be achieved at high adjustment speeds using electronic devices, but electronics have their price.

With increasing demands for accuracy, there was a consequent increase in demand for security and productivity in measurement processes. A good example of the continuous progress towards increased accuracies, greater reliability and higher operating speeds in length and angle measuring systems is provided by numerically-controlled machine tools. Up to about 1970, a resolution of 0.01 mm was considered adequate, whereas a resolution of 0.001 mm has been standard since 1980. Today digital length comparators work to resolutions of 0.001 mm and laser interferometers discriminate up to 0.00001 mm.

Measuring systems with a resolution of $0.02 \mu m$ are required and are available to produce highly integrated semiconductor memories (work is currently being done as a result of the development of 64 Mbit memories). When we talk about increasing power in measurement processes, what we mean essentially is resolution, accuracy, speed of measurement and, last but not least, reliability. Another development made possible in the last few years by the microprocessor is intelligent sensors, in which data processing can take place in the sensor, and which are capable of communicating with a central control via a databus.

This volume is an attempt to bring together all the main principles of measurement in current use and to provide a simple comparison of their characteristics. Moreover, there are numerous processes which had disappeared from current use for any one of a number of reasons, but which have nevertheless been described in some detail because of their technology which is interesting in itself.

Nowadays electro-optical processes, in which a measurement base in the form of an incremental grid or code is scanned photoelectronically, are dominant. For a long time there were two different schools of thought, one in favour of the incremental method of measuring and the other in favour of the absolute method of measuring. This was essentially due to the fact that the reliability of incremental systems was dubious, which in turn was due to the difficulty of detecting false counts caused by interference with or absence of pulses. Today this dispute has been settled.

As incremental measurement processes, due to their now proven reliability and, more especially, their cost advantages, are used more frequently today than absolute systems, they are treated in the same depth.

The ease with which trends in engineering can be reversed is shown by the fact that absolute systems are being used again, not principally for reasons of reliability, but because it is not necessary to set up a datum point even after a shut-down or power failure, and in this way time is saved. It is still mandatory to select an absolute distance or angle measuring instrument if manual adjustments could be made on a machine when idle which could lead to a dangerous situation when it was switched on again.

Analogue length and angle measuring instruments, such as resolvers and inductosyns, can fully match digital instruments in terms of resolution and accuracy. The output signals of these analogue instruments can be very accurately digitized and converted to a suitable display for electronic data processing. As this means they are no longer any different from purely digital positional instruments as far as data transfer and data processing are concerned, all systems with analogue instruments which are fitted with typical analogue-to-digital converters are included under the general heading of digital position-measuring processes. The advantages of digital measurement apply in the same way to these hybrid systems.

The accuracy of a measurement depends solely on the accuracy of the measurement process and not the accuracy of the transfer or readout. This means that remote measurement, measurement in inaccessible locations, central indication and automatic recording are possible without loss of accuracy. Besides the characteristics of the measurement processes themselves, the various metrological methods are shown in conjunction with mechanical connecting links, as false conclusions could be drawn from the results of a measurement if the type of connection is not known.

Irrespective of the physical principal of the measurement base used and the relative sampling method, certain processes for signal analysis have developed which are used in many, often very different, measuring instruments. For this reason, the description of these processes is interesting apart from the specific application. For example, these include all processes in which two signals are produced which follow the sine or the cosine of the quantity to be measured. Examples are high-resolution incremental photoelectric distance measurement processes, inductive instruments such as resolvers or inductosyns and various capacitive or magnetic processes. The positional information is then contained either in the ratio of two amplitude values or in the difference between two phase angles. In both cases very high resolutions can be obtained provided the signals are to a large extent free of harmonic waves.

A basic principle equally widespread throughout very different applications is that of distance measurement by means of measuring pulse timing. Both the ultrasonic method and the radar distance measurement process, as well as certain processes in laser distance measuring, must be mentioned here. Another important group consists of the interferometers which can be used for both distance and angle measurement (laser and fibre-optic gyros).

The use of a specific principle of measurement for a specific application depends not only on the accuracy and resolution required but also on additional requirements such as permissible temperature range, resistance to penetration of dust or liquids and insensitivity to shock and vibration stress.

We shall therefore have to consider each principle of measurement to be used in the light of such additional requirements and not just from the point of view of absolute accuracy. The choice between purely analogue measurement, the use of analogue measuring instruments with downstream analogue-to-digital converters and purely digital measurement processes is made only in a particular case after all the requirements have been taken into account.

# 2

## Principles

## 2.1 Measurement technology

### 2.1.1 Units

The following is used as a physical equation in DIN 1313: quantity = numerical value × unit (e.g. $s = 1.824\,\text{m}$). On 3 July 1970, the German law on units of measurement, which states that henceforth only the fundamental quantities specified in the SI (Système International d'Unités) international system of units and the units derived from them may be used in official and commercial transactions, came into effect.

The seven fundamental units of measurement are:

the kilogram (kg) for mass;
the metre (m) for length;
the second (s) for time;
the ampere (A) for electric current;
the kelvin (K) for thermodynamic temperature or kelvin scale. The Celsius scale, symbol $t$, is defined by the expression $t = T - T_0$ where $T_0 = 273.15$ K and $T$ is the temperature on the kelvin scale at the time of measurement;
the candela (cd) for luminous intensity;
the mole (mol) for amount of substance.

The major derived units, some of which have been amended by law, are as follows:

- force: the newton (N)
  $1\,\text{N} = 1\,\text{kg} \cdot \text{m}\,\text{s}^{-2}$;

- work, energy, quantity of heat: the joule (J)

$$1\,J = 1\,Ws = 1\,N\cdot m = \frac{1\,kg\cdot m^2}{s^2}$$

- power: the watt (W)

$$1\,W = \frac{1\,kg\cdot m^2}{s^3};$$

- pressure

$$\frac{N}{m^2} = 10^{-5}\,bar;$$

- magnetic flux: the weber (Wb)

$$1\,Wb = 1\,Vs = \frac{1\,Ws}{A};$$

- magnetic flux density (inductance): the tesla (T)

$$1\,T = \frac{1\,Wb}{m^2} = \frac{1\,Vs}{m^2};$$

- magnetic field strength: amperes per metre $(A\,m^{-1})$[1;2]

### 2.1.2   Errors of measurement and accuracy

When working with error quantities attention must be paid to the preceding sign.

The error:

$$F = A - W$$

is described as positive if the uncorrected (wrong) value, the indication $A$, is greater than the true value $W$, thus error = false – true.

On the other hand the correction has the opposite preceding sign:

$$K = -F = W - A.$$

The relative error is:

$$f = \frac{A - W}{W} = \frac{E}{W} = \frac{-K}{W},$$

thus relative to the true value. In the case of small errors ($f \ll 1$) a low-level error occurs if $W$ is replaced by $A$ in the denominator, which is often more convenient when making evaluations. Percentage errors are described as 100 times $f$ [3]:

$$p = 100\,f$$

A clear distinction must be made between systematic and random errors when considering errors and where necessary error correction. Systematic errors are caused by the method of measurement, environmental influences and the observer's personal qualities.

Examples of the above are:

- inaccuracies in measurement base, measuring instruments and methods of measurement,
- environmental influences such as temperature, air pressure, humidity, tension, frequency, external electrical and magnetic fields,
- effects of the observer, such as attentiveness, practice, eyesight, power of judgement (of lesser importance in digital methods of measurement).

Systematic errors may, if known, be compensated by applying corrections. If systematic errors cannot be easily detected then at least one estimate is required to enable the uncertainty of measurement to be defined. Random errors are caused by unpredictable variations in the measurement base, measuring instruments (e.g. friction), measured object, environment and observers which take place during measurement. Unlike systematic errors they can be avoided by taking a large number of measurements and obtaining a mean value.

Statistical methods can be used for the computation of errors [4;5;6] where these are random errors. The prerequisite for this is that the frequency $y$ of the occurrence of certain sample values $x$ (distribution density of $x$) should produce an approximate normal distribution (Gaussian distribution curve), Figure 2.1. The expression for this curve is:

$$y = \frac{1}{\sigma \sqrt{2\pi}} e^{(x-\bar{x})^2/2\sigma^2} \tag{1}$$

Figure 2.1 shows three areas of normal distribution and the percentages for the number of sample values which can be expected within these areas. Thus 99.7% of all sample values can be expected in the area $\bar{x} \pm 3\sigma$. The quantity $\sigma$ is referred to as the standard deviation. It is the most significant operand for the random deviations of the individual values from their mean.

The mean value (arithmetic mean) of a series of measurements of $n$ separate sample values $x_1, x_2 \ldots x_n$ is:

$$\bar{x} = \frac{1}{n} \sum_{i=1}^{n} x_i \tag{2}$$

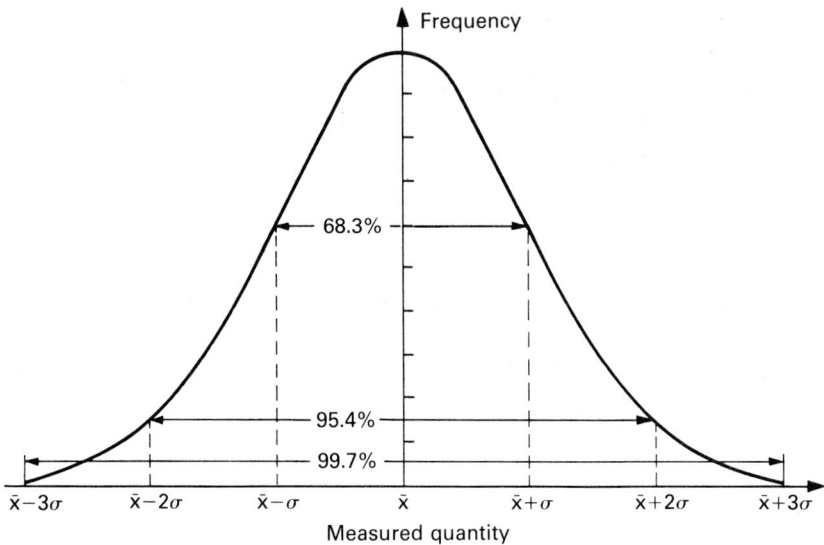

Figure 2.1    Normal distribution (Gaussian curve).

This mean value is the most probable value for the true value. The probability of $\bar{x}$ being the same as the true value depends on the number $n$ of individual measurements. As a function of $n$ and the statistical certainty, two limit values, above and below the mean, between which the true value lies can be determined. These limits are referred to as the confidence level of the mean. The standard deviation is defined as the mean square error of the individual observations:

$$\sigma = + \sqrt{\frac{1}{n-1} \sum_{i=1}^{n} (x_1 - \bar{x})^2} \qquad (3)$$

For very large $n$ ($n \geqslant 200$), $\sigma$ is the standard deviation of the population. For smaller values of $n$ the letter $s$ replaces $\sigma$ to denote the fact that the radicand is not an assertion of the population. In order to calculate the limits of confidence from $s$, an additional factor (Student's $t$ distribution) must be taken into consideration.

The following expression applies for the upper confidence limit:

$$\bar{x} + \frac{t}{\sqrt{n}} s \qquad (4)$$

and for the lower confidence limit:

$$\bar{x} - \frac{t}{\sqrt{n}}\, s \qquad\qquad (5)$$

The values [4;5] as listed in Table 2.1 therefore apply to $t$ and $t/n$ as statistical certainties ($P = 95\%$ and $P = 99\%$).

Where $n \geqslant 200$, $t$ is constant and $s = \sigma$.

Table 2.1   Factor $t$ for statistical certainty $P = 95\%$ and $P = 99\%$

| | $P = 95\%$ | | $P = 99\%$ | |
|---|---|---|---|---|
| $n$ | $t$ | $t/\sqrt{n}$ | $t$ | $t/\sqrt{n}$ |
| 3 | 4.3 | 2.5 | 9.9 | 5.7 |
| 4 | 3.2 | 1.6 | 5.8 | 2.9 |
| 5 | 2.8 | 1.24 | 4.6 | 2.1 |
| 6 | 2.6 | 1.05 | 4.0 | 1.6 |
| 8 | 2.4 | 0.84 | 3.5 | 1.24 |
| 10 | 2.3 | 0.72 | 3.25 | 1.03 |
| 20 | 2.1 | 0.47 | 2.9 | 0.64 |
| 50 | 2.0 | 0.28 | 2.7 | 0.38 |
| 200 | 1.97 | 0.14 | 2.6 | 0.18 |
| >200 | 1.96 | 0 | 2.58 | 0 |

The expressions for the confidence limits are:

$$\bar{x} \pm \frac{1.968\sigma}{\sqrt{2}} \text{ for statistical certainty } P = 95\% \qquad (6)$$

$$\bar{x} \pm \frac{2.588\sigma}{\sqrt{2}} \text{ for statistical certainty } P = 99\% \qquad (7)$$

Accordingly, the result of a measurement which is the end result of a series of measurements of $n$ individual values will be indicated as mean value $\bar{x}$ together with the level of confidence for a specific statistical certainty. Systematic errors are taken into consideration in these indications only if $\bar{x}$ has already been corrected by the amount of the systematic errors detected. If, in addition, systematic errors which are undetectable and can therefore only be estimated are to be taken into consideration, then the uncertainty of measurement is indicated instead of the error of measurement. The end result is then:

$$y = \bar{x} \pm u.$$

The overall uncertainty of measurement is:

$$u = \left| \frac{t}{\sqrt{n}}\, s \right| + |f| \qquad\qquad (8)$$

where $f$ is the sum of the estimated systematic errors which are undetectable.

These relations apply where an error of a single measured quantity is being considered. If, however, the result of a measurement is a function of several measured quantities (e.g. the resistance value as the result of a current or voltage measurement) error propagation must be taken into account when determining the error of measurement. Let the separate measured quantities be $x_1$, $x_2$ ... $x_v$ and the errors determined in the way described above be $\Delta x_1$, $\Delta x_2$ ... $\Delta x_v$. What we are looking for is the error of dimension $y$, which according to formula is a function of $x_1$, $x_2$ ... $x_v$: $y = f(x_1, x_2 ... x_v)$. Provided that the errors of the separate variables in this equation are small, the expression for systematic errors is:

$$dy = \frac{\partial y}{\partial x_1} \Delta x_1 \; \frac{\partial y}{\partial x_2} \Delta x_2 + \ldots \frac{\partial y}{\partial x_v} = \sum_{i=1}^{v} \frac{\partial y}{\partial x_i} \Delta x_i$$

which, for example, produces:

$$y = a x_1^n x_2^m$$

$$dy = n \frac{y}{x_1} \Delta x_1 + m \frac{y}{x_2} \Delta x_2$$

When considering random errors the statistically separate variables and their standard deviations $s_1$, $s_2$ ... $s_v$ must be known (obtained from a series of measurements with an equal number of individual values).

Additionally, the measured values must correspond to a normal distribution and be $s_i \ll \bar{x}_i$. The standard deviation $s_y$ of the function $y = f(\bar{x}_1, \bar{x}_2 ... \bar{x}_v)$ can then be determined from:

$$s_y = \sqrt{\sum_{i=1}^{v} \left( \frac{\partial y}{\partial \bar{x}_i} s_i \right)^2} \tag{9}$$

The Gaussian law of error propagation gives the quadratic mean value of random errors.

Example:

$$y = m\bar{x}_1 + n\bar{x}_2$$

$$sy = \sqrt{(ms_1)^2 + (ns_s)^2}$$

Strictly speaking, these relations only apply to population standard deviations $\sigma_i$. In samples with the same number $n$ of separate individual values (equal statistical certainty) the confidence level of $y$ is calculated from [4;5;7]:

$$\frac{t}{\sqrt{n}} s_y = \sqrt{\sum_{i=1}^{v} \left( \frac{\partial y}{\partial \bar{x}_i} \frac{t}{\sqrt{n}} s_i \right)^2} \tag{10}$$

## 2.2 Terms used in information technology

According to DIN 44300 [8], a message is characterized by digits or continuous functions which for the purposes of transmission display information on the basis of known or assumed stipulations. The term 'information' is not defined further in this context but is used as in everyday speech to mean knowledge of facts and events. The message is produced by using an information source to make a selection from a series of predetermined options in a set of characters. This character set can be an alphabet, a language or a series of numbers. If we confine ourselves to a specific finite series of numbers which has been uniquely allocated to specific items of information by agreement between the sender and the recipient, then we are talking about a code. Thus, for example, the allocation of the on and off conditions of a switch in an electrical circuit to two logical conditions is a coding, if for example logical condition 1 is allocated to the off condition and logical condition 0 to the on condition of the switch. This type of coding with only two digits is referred to as binary coding (binary code).

Technically, binary coding is easy to put into practice, as only two clearly defined conditions have to be created. Examples are: current on/off, voltage on/off, light on/off, pressure on/off, positive/negative polarity, magnetic north/south. A choice between two characters is also the simplest possible decision. The choice of an option from $N$ options in a character set $X$ can also be brought back to a simple Yes/No decision. The number of binary decisions necessary until the final result is obtained is the decision content of the character set and is given in bits. If the size of a character set $X$ is $N = 2^n$, then $n$ is the decision content. It is produced as a logarithm to the base of 2 of the number $N$; $n = \log_2 N = 1bN = H_0(X)$. In relation to the character set $X$ the decision content is given as $H_0(X)$. Let us take as an example of choosing a character from a character set the well-known brainteaser, how many weighing operations does it take using one pair of scales to find out which one of twelve balls of equal external dimensions weighs less than the others. As a first step six balls are placed in each scale and the six heavier ones discarded (first decision). Of the six lighter balls, three are placed in each scale and again the heavier ones are discarded (second decision). Of the remaining three balls two are placed on the scales to determine whether there is equilibrium or not (third decision). If there is equilibrium, it means that the third ball, not on the scale, is the one we are looking for (fourth decision). With twelve characters the decision content is lb 12 = 3.58. If we round this up to the next whole number we get the required number of binary decisions (four), which corresponds to the above example for identifying a character. For a recipient who is to receive a specific message the decision content meets the technical requirement provided it lays down the maximum number of binary digits required. It is therefore important to consider to what extent a message constitutes an item of information. Although a wire carrying a constant voltage is transmitting a message, e.g. logical value 1, this message has no information content. The key criterion for information content of a message is the probability of the occurrence of the message. The probability of the occurrence of a result $x_j$ is $p(x_j)$. The range of $p(x_j)$ lies between 0 and 1:

$$0 \leqslant p(x_j) \leqslant 1 \tag{11}$$

and the expression:

$$\sum_j p(x_j) = 1 \tag{12}$$

also applies.

Taking a dice as an example, this means that the probability of throwing a specific number in one throw is equal to 1/6. The probability of throwing any number is equal to the sum of the individual decisions, thus $6 \times 1/6 = 1$.

In principle it is true that the information content of a digit is greater, the more improbable its occurrence. An event which is expected and the occurrence of which is almost certain has no information content. Thus, after it has been raining for a week, the forecast that it will rain again the next day has no significant information content, whereas in the same circumstances the forecast that the sun will shine has a considerable information content. The information content $I(x_j)$ is defined as:

$$I(x_j) = \text{lb} \frac{1}{p(x_j)} = -\text{lb} p(x_j).$$

In a character set the average information content $H(X)$ is described as entropy.

The following expression applies:

$$H(X) = \sum_j p(x_j)I(x_j) = -\sum_j p(x_j)lbp(x_j) \tag{13}$$

The average information content is an important characteristic of an information source. It is thus a maximum and equal to the decision content if $p = 1/N$, or all signs have the same probability of occurring:

$$H_{max} = H_0 = lbN \tag{14}$$

This difference between the maximum value and all the possible smaller values is described as redundancy:

$$R = H_0(X) - H(X) \tag{15}$$

In this case redundancy means having an ample sufficiency. Redundancy arises, for example, if not all the digits in a set of digits are used. This is shown in the example of the BCD Code (BCD, Binary-Coded Decimals), Table 2.2. The BCD Code is used for the binary sub-coding of decimal places. Four binary digits (bits) provide 16 combinations, only ten of which are required. The remaining six are described as pseudotetrads. Which six combinations are to be defined as pseudotetrads can be decided at random. In the excess-three code,

Table 2.2   BCD code

| Decimal number | BCD-Code | | | | | Test bit |
|---|---|---|---|---|---|---|
| | $2^3$ | $2^2$ | $2^1$ | $2^0$ | | |
| 0 | 0 | 0 | 0 | 0 | | 1 |
| 1 | 0 | 0 | 0 | 1 | | 0 |
| 2 | 0 | 0 | 1 | 0 | | 0 |
| 3 | 0 | 0 | 1 | 1 | | 1 |
| 4 | 0 | 1 | 0 | 0 | | 0 |
| 5 | 0 | 1 | 0 | 1 | | 1 |
| 6 | 0 | 1 | 1 | 0 | | 1 |
| 7 | 0 | 1 | 1 | 1 | | 0 |
| 8 | 1 | 0 | 0 | 0 | | 0 |
| 9 | 1 | 0 | 0 | 1 | | 1 |
| – | 1 | 0 | 1 | 0 | | 1 |
| – | 1 | 0 | 1 | 1 | | 0 |
| – | 1 | 1 | 0 | 0 | Pseudotetrads | 1 |
| – | 1 | 1 | 0 | 1 | | 0 |
| – | 1 | 1 | 1 | 0 | | 0 |
| – | 1 | 1 | 1 | 1 | | 1 |

for example, these are the first and last three tetrads. The decision content of the 16 digits is

$$H_0(X) = 1bN = 1b\ 16 = 4\ \text{bits}.$$

The probabilities for the occurrence of one of the first ten digits are equal and therefore $p(x_j) = 1/10$ and those for the next six digits exactly zero. The information content of one of the first ten digits is $I(x_j) = 1b(p(x_j) = 1b\ 0.1$. Thus the average information content of the BCD Code, the entropy, is:

$$H(X) = \sum_j p(x_j)I(x_j) = \sum_1^{10} 0.1\ lb\ 10 = lb\ 10 \qquad (16)$$

The redundancy is therefore

$$R = 1b\ 16 - 1b\ 10 = 4 - 3.32 = 0.68\ \text{bits}.$$

Redundancy, however, also arises if the occurrence of a digit is dependent on a previous one. The entropy of a pair of digits is calculated as follows:

$$\begin{aligned} H(X\ Y) &= \sum_j \sum_k p(x_{jy_k})I(x_j)y_k) \\ &= H(X) + H(Y/X) \\ &= H(X) + H(Y;X) \end{aligned} \qquad (17)$$

If the event spaces $X$ and $Y$ are independent of each other, the expression $H(X;Y)$, the synentropy, becomes equal to zero. If, however, this is not the case, the probability of the occurrence of a digit depends on the preceding digit and the entropy of the pair of digits is reduced by the synentropy which is therefore also the redundancy. If we again consider the BCD Code and ask about the synentropy in two adjacent codewords, we have to ask about the probability of the occurrence of the second code word, assuming the first one has already occurred. For example, the probability of the occurrence of 0111, if 0110 was already there, is exactly 50%. According to equations (16) and (17), the expression for the synentropy is

$$\begin{aligned} (X;Y) &= H(Y) - H(X/Y) = 0.11b\ 10 - 0.05\ 1b\ 20 \\ &= 0.332 - 0.216 = 0.116. \end{aligned}$$

The synentropy in two adjacent words in the BCD Code is therefore 0.116 bits.

An example of the desired redundancy is what is known as the parity bit which can, for example, be set so as to increase the number of uneven logic-1 bits in a codeword (Table 2.2). In a receiving device transmission errors can be recognized in a digit position (simple error) by comparing the sum of digits (mod 2) with the parity bit. Although, therefore, redundancy increases the hardware requirement, it also provides greater security. Redundancy is likewise required when a message is transmitted twice for security reasons.

## 2.3 Methods of measurements

If we are given the task of measuring the geometric sizes of angle or length the first thing we think of is a direct comparison of these sizes with an appropriate standard or *étalon*. In the case of angles such an *étalon* could be an exact circular graduation with markings at set distances against which angle values are given (protractor). The same is true for the standard of length (e.g. a metre rod). Such measurements are described as direct or absolute, as the measured size and the standard are compared directly, the standard having a fixed zero point and supplying absolute numerical data on the measured value. At the same time this method of measurement involves the digitizing of the measured size as the standard divides the analogue physical size into increments, e.g. into millimetres or degrees, and supplies the measured value as the number of increments corresponding to the measured size. Besides the length or angle measurement there is often a requirement for a position to be determined. When starting up a machine tool, for example, it may be necessary to begin or continue work on a workpiece from a specific position. Absolute measurement bases are also suitable for this task as they identify each position with a corresponding code, e.g. with unique numbers. The counterpart of absolute measurement is relative or incremental measurement. In this method the number of increments on a grid (a circular dial or a linear scale) which correspond to the measurand are counted. The starting point for counting, i.e. the zero point on the grid, can be selected at random as no point is distinguished from another. In practice, however, incremental electrical position measurement processes are usually provided with an additional zero marker so that the counters connected to them can be set to zero at a specified position and determination of positions can be carried out in addition to the measurement of distances or angles.

It will be seen when comparing absolute and relative position measurement that both methods have their advantages and disadvantages. Absolute measurement offers greater security against measurement and transmission errors, especially if error detection is made possible by means of additional test signals (parity bits). A further advantage is the fixed zero point of the absolute system, which can be found again irrespective of shut-downs or machine failure.

These advantages are gained by greater complexity and therefore at higher costs. In the case of an absolute measurement process zero displacement is only possible by means of an arithmetic unit which adds the amount of the zero displacement to, or subtracts from, the measured value.

Incremental systems are characterized by lower cost of detection and transmission combined with simple zero selection and displacement. They do, however, have the disadvantage that measurement errors due to interfering pulses or miscounting cannot be detected. In the event of power failure or when the measurement process is restarted no information on current position is available. Measurement must be started all over again from a reference point.

Measurement processes in which a small absolute range is repeated cyclically are often described in the literature as cyclically absolute systems. An example of this is an angle encoder with a single disc which is used for measuring angles of more than 360°, thus of more than one revolution. The number of revolutions must be counted and therefore measured incrementally. Other examples are inductive measurement processes such as the inductosyn (see Section 3.13), in which an analogue range is repeated at intervals of 2 mm. Strictly speaking, in these cases it is a question of incremental systems with interpolation of the individual increments.

Although direct comparison of the measurand with a standard is an obvious method of measurement it is not the only one. Thus a distance measurement can be made by converting longitudinal motion into rotation by means of a measuring spindle and screw nut or using a rack and pinion, Figure 2.2.

In many cases longitudinal motion is discontinued through rotation in any event if, for example, electric motors are used as a drive mechanism. A further example is the series connection of measuring mechanisms. Indirect measurement is widespread, as it is often associated with design or even cost advantages. At all events,

Figure 2.2 (*Opposite*)   Position sensing system. (a) Direct linear measurement. (b) Indirect linear measurement using rack-and-pinion. (c) Indirect linear measurement using measuring spindle. (d) Direct measurement of angles.

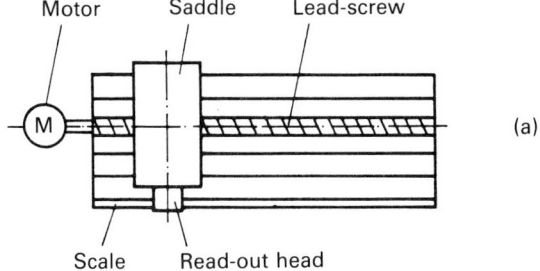

(a)

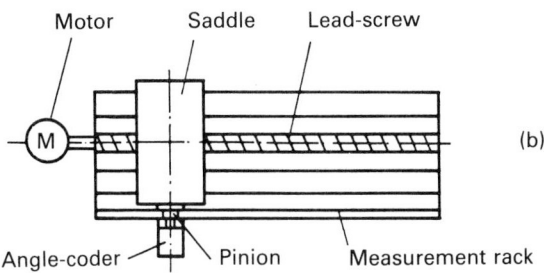

(b)

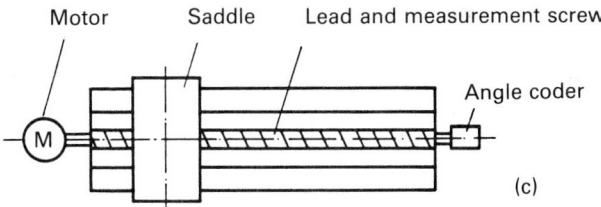

(c)

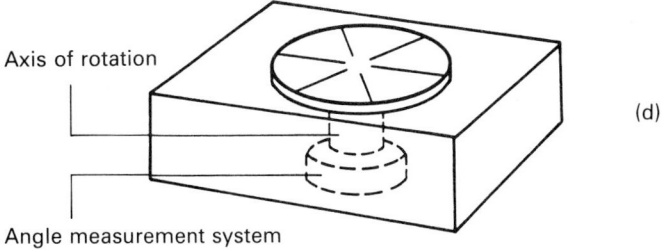

(d)

errors in the mechanical connecting links have an effect on the accuracy of measurement.

In the case of machine tools the terms direct and indirect measurement require additional clarification, as it is only very rarely that the workpiece itself is actually measured. Such measurement is possible only in special cases, since chips, oil, coolant and complicated workpiece shapes prevent direct measurement, always assuming any suitable methods of measurement exist.

Consequently, slide paths are measured instead of the workpiece itself. For this it is essential that both the workpiece and the tool are held firmly in position in their clamping or chucking device and do not deviate from it significantly throughout the machining process. Deformations of the workpiece, the tool or machine components affect the accuracy of measurement and must therefore be kept within the narrowest possible limits. In this connection direct distance measurement always refers solely to the direct measurement of moving machine components.

In the production and testing of precision measures, direct comparison is made with a measurement standard. This should be done in accordance with Abbé's comparator theory which results in only secondary errors of measurement. The test specimen and standard are aligned, one behind the other, on a sliding table (first principle of measurement engineering). A comparison is made using two fixed microscopes which are kept a constant distance apart. It is important that the table and not the microscopes be moved during measurement (second principle of measurement engineering). Attention to these two principles helps to avoid primary errors of measurement due to play and irregularities in the slideways, which cause a certain amount of tilting and twisting. If the specimen and standard are arranged parallel to each other at distance $a$ (breach of the first principle of measurement engineering) rotating the measuring table by angle $\alpha$ will result in an error $f_1 = a$ arc $\alpha$. If the second principle of measurement engineering is breached, tilting of a microscope by angle $\beta$ at height $h$ results in error $f_2 = h$ arc $\alpha$. Both of these are primary errors as they depend linearly on $\alpha$ and $\beta$, respectively. If we take as a comparison the effect of a tilting of the comparator by $\alpha$ in the longitudinal comparator, in which the specimen and standard are placed at distance $l$ one behind the other, then the errors of measurement is $\Delta l = -1/2$ arc$^2 \alpha$. A secondary error does not usually affect the result of a measurement [9;10].

# 3

## Industrial sensors and measurement systems

### 3.1 Analogue angle and displacement sensors

Analogue angle and displacement sensors often have cost advantages over digital sensors and give absolute measurements in every case— unlike, for example, incremental digital sensors. They have, at least in theory, an infinitely high resolution and under specific conditions they can provide shorter sensing times. If the power limits of analogue and digital measurement processes are compared a reasonable maximum number of amplitude increments $m$ can be estimated in accordance with the following relation [11]:

$$m = 1 + \frac{1}{2F} \tag{18}$$

In this $F$ is a given statistical error. Increasing the resolution by refining the number of amplitude increments only makes sense so long as the associated quantization does not exceed the size of the error. If the statistical error in a sensor amounts to 0.5% then a maximum of 101 increments is useful.

The relation:

$$f_0 = \frac{1}{2t_E} \tag{19}$$

can be used to describe dynamic characteristics. In this $f_0$ is the band limit and $t_E$ the response time of an ideal rectangular pulse. If the same observations are now assumed for digital measurement processes, then the following applies in the static case:

$$m = 1 + fT \qquad (20)$$

for the number of amplitude increments, where $f$ is the counting frequency and $T$ the measuring time.

This allows for the fact that an uncertainty of measurement of one increment may occur during counting. Equation (20) can be considered to be a fundamental principle of digital metrology. During a counting operation the measuring time is 1 s and the frequency 1 kHz. This produces 1001 distinguishable increments, so that $f$ is the maximum frequency at which sample values can be gathered.

Dynamic characteristics are described by the relation:

$$f_g = \frac{1}{2T} \qquad (21)$$

where $f_g$ is the limit frequency and $T$ is the measuring time. In digital systems the linking of the dynamic and the static characteristics can be described using equations (20) and (21):

$$m = 1 + \frac{f}{2f_g} \qquad (22)$$

There is now a firm connection between $m$ and $f_g$. This means that maximum sensing speed and obtainable resolution cannot be altered independently of each other. A high resolution requires a low limit frequency and vice versa.

This is a significant disadvantage of digital processes as against analogue processes, in which it is possible to have high resolution and high measuring speed at the same time.

### 3.1.1   Position indication using gravity

In order to gauge the position of machine parts which can be adjusted on horizontal spindles, such as limit stops on saws or presses, the setting of gaps between rollers, control mechanisms, etc., position indications are used which are built directly into the hand-wheel required for making adjustments, Figure 3.1 The principle on which the position indication operates is that a weight is suspended like a pendulum in a cylindrical housing which turns as the hand-wheel is adjusted. When the housing is rotated the weight remains at rest owing to gravity. The relative motion between the housing and the weight is transmitted from a pinion firmly mounted on the housing to a mechanism located on the body of the weight and from there in turn to one or, if desired, two differently geared down pointers, Figure 3.2. The selection of the transmission ratios is made without regard to the measuring task (pitch of spindle, range of measurement). The main features of this principle of measurement are its simplicity, reliability and robustness. Where a special risk of corrosion is present, or strong vibrations and shocks are expected, the device may be oil-filled. The resolutions which can be achieved depend on the pitch of the spindle and may reach 0.01 mm. The uncertainty of measurement is about 0.05 revolutions.

Figure 3.1  Position indication in spoked handwheel. (Picture courtesy of Gebrüder Lohmann KG.)

## 3.1.2  Resistance transducers

The resistance value $R$ in ohms of a metal conductor is derived from the relation:

$$R = \rho \frac{l}{A} \tag{23}$$

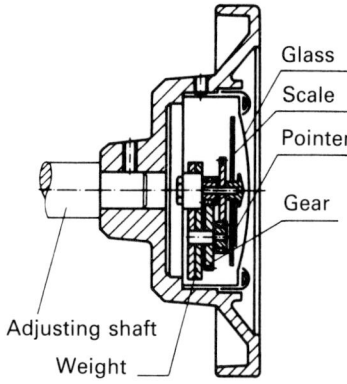

Figure 3.2   Cross-section of handwheel showing position indication.

Here $\rho$ is the specific resistance in $\Omega\,$m, $l$ the length in m and $A$ the cross-sectional area in m$^2$.

In the construction of resistance transducers, each of these three quantities may be altered. In potentiometers, the transducers most frequently used on a resistance basis, either the length is varied and a quantity derived from it, e.g. current, is measured, or the dividing ratio is displayed. The second case is more common as the absolute resistance quantity has no direct influence on the result of a measurement. Each potentiometer unavoidably contains inductive and capacitive components. These can, however, be ignored when using direct current or low-frequency alternating current. A potentiometer consists basically of a resistance transducer with a movable contact. Depending on its construction, the contact can be moved on a straight or arc-shaped path, Figure 3.3. Consequently, potentiometers are suitable for measuring displacements or angles. A common type of construction uses a spindle drive to move the slide.

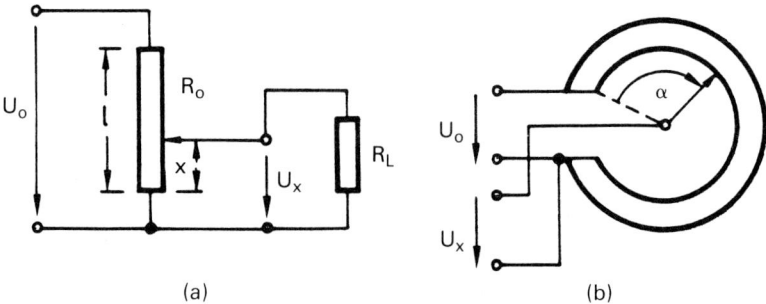

Figure 3.3   Linear (a) and amular (b) potentiometer.

The actual resistance transducer consists either of a wire-encased resistor, or a film resistor with a layer of carbon or conductive plastics. Magnetic field resistors which operate without contact and are therefore free of friction are a more recent development. The main characteristic quantities of each potentiometer are resistance, permissible power loss, tolerance, linearity and resolution. Typically the resistance is between $100\,\Omega$ and $100\,k\Omega$. Power loss will increase at a given voltage where resistance is low. At higher resistances, not only do the inductive and capacitive components play a greater part, but there is also an increased risk of electromagnetic disturbances. A compromise must therefore be reached between the various requirements in each individual case.

The maximum supply voltage $U_{0\,max}$ of a potentiometer is derived from:

$$U_{0\,max} = \sqrt{P_{zul}\,R}, \text{ [V]} \tag{24}$$

where $P_{zul}$ is the permissible power loss in W, $R$ is the resistance value in $\Omega$.

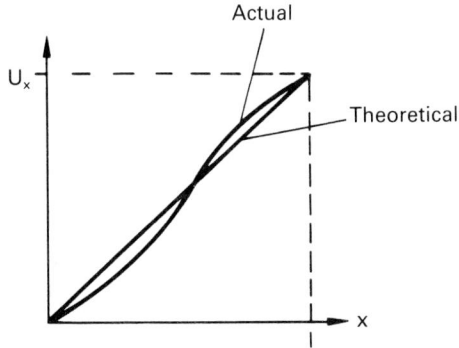

Figure 3.4   Linearity deviation in potentiometers due to manufacturing tolerances.

The tolerance of the potentiometer is the difference between its actual and nominal resistance, expressed as a percentage of nominal. It lies between $\pm1\%$ and $\pm10\%$. Typical values are between $\pm3\%$ or $\pm5\%$. Linearity is a very important characteristic quantity. Even in the no-load condition when, for example, the output voltage is being determined by a compensation method, the output voltage does not change strictly linearly with the slideway or the angle of rotation of the slide, Figure 3.4. The deviation of the actual resistance from the ideal curve is normally between 0.05% and 1%. Very good results are obtained using wire-wound potentiometers, but these have the disadvantage that the resistance changes sharply from winding to winding. In practice, up to 25 windings per mm can be produced. This means that in linear potentiometers the resolution is restricted to about 40 $\mu$m. The resolution of a single-gang ring potentiometer with a diameter of 25 mm is about 0.2°. Potentiometers of the carbon film or electrically conducting plastics type, on the other hand, have a theoretically infinite resolution. In practice, 0.01 mm is achieved in displacement measurement and 0.01° in angle measurement. Their linearity, however, is usually inferior to that of wire potentiometers. In addition, it is advisable to restrict the slide current to a maximum of 1 mA, as otherwise working life is seriously reduced.

These difficulties are largely avoided by a combination of both methods: a wire resistance element is coated with electrically conducting plastics on the side on which the slide is situated. The wire-wrapping technique provides good linearity while the electrically-conducting plastics film ensures infinitely fine resolution. In addition, the slide can be given a high contact pressure, thus ensuring not only a low transition resistance but good vibration and shock characteristics as well [12].

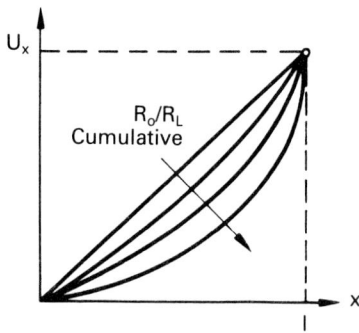

Figure 3.5   Linearity deviation in potentiometers due to load resistance.

As previously mentioned, the linearity of displacement or angle measurement not only depends on the linearity of the potentiometer itself, but on the load as well. The output of the potentiometer is loaded by the current demand of the downstream circuit element. Figure 3.5 shows how an increasingly non-linear curve develops from a linear curve of the output voltage with a reduction of the load resistance $R_L$ (Figure 3.3a). The equation is:

$$\frac{U_x}{U_o} = \frac{1}{\dfrac{l}{x} + \dfrac{R_o}{R_L}\left(1 - \dfrac{x}{l}\right)} \tag{25}$$

In the ideal case of the unloaded potentiometer ($R_L = \infty$), this becomes:

$$\frac{U_x}{U_o} = \frac{x}{l} \tag{26}$$

If $R_L = R_0$, the maximum error is 12% of the end scale value. It falls to 1.5% when $R_L = 10\,R_0$.

The creation of high load resistances presents no problems if operational amplifiers are used as impedance transformers. With linear potentiometers the range of measurement which can be achieved depends on the length of the resistance element. In practice, most linear potentiometers are between 25 mm and 300 mm long, but lengths of up to 2000 mm are also available. In the case of rotary potentiometers standard products are designed for one to ten rotations.

Series connection mechanisms are available for adapting to a given measurement task.

There may be circumstances where it is necessary for a very small angular movement to be geared up or a measurement range of considerably more than ten rotations to be adapted for a ten-gang potentiometer. Figure 3.6 shows an example of a potentiometer adapted to meet tough environmental conditions. A robust steel housing, a ball-bearing shaft and an internal slip clutch make it suitable as a transducer of angles or displacements in engineering.

### 3.1.3   *Capacitive transducers*

The capacitance of a plate capacitor is proportional to plate surface. $A$ and inversely proportional to plate gap $d$:

$$C = \frac{\varepsilon_0 \varepsilon_r\, A}{d}\, [\text{F}] \tag{27}$$

where $\varepsilon_0$ is the permittivity of empty space and $\varepsilon_r$ the relative permittivity ($\varepsilon_0 = 8.854\ 10^{-12}\ [\text{F}\,\text{m}^{-1}]$; $\varepsilon_r = 1$ in air).

Figure 3.6   Geared potentiometer. (Picture courtesy of Erwin Halstrup Multur GmbH).

The capacitance is therefore a function of the geometrical arrangement and the dielectric used. For a concentric arrangement of length *l* (cylindrical capacitor) the equation is:

$$C = \varepsilon_o \varepsilon_r \frac{2\pi l}{ln\frac{r_a}{r_b}} \tag{28}$$

where $r_a$ is the radius of the outer cylinder and $r_b$ is the radius of the inner cylinder.

The use of differential capacitors is recommended for measuring purposes as these to a large extent avoid the non-homogeneity of field at the edges of the electrodes, Figures 3.7 and 3.8. In practice, the capacities used in angle and distance measurement vary between 1 and 500 pF. In order to prevent the impedance from becoming too great, suitable supply frequencies between 0.1 and 1 MHz are selected. In addition, the insulation resistance selected must be high enough to keep shunt errors to a minimum. This requirement is difficult to adhere to when measurements are to be carried out at high humidity.

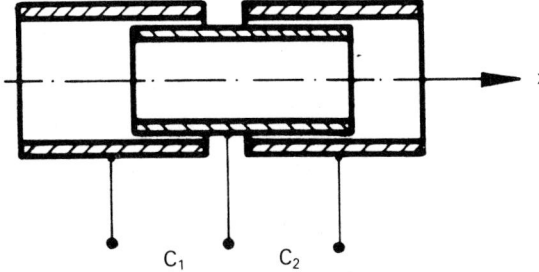

Figure 3.7   Differential cylinder capacitor.

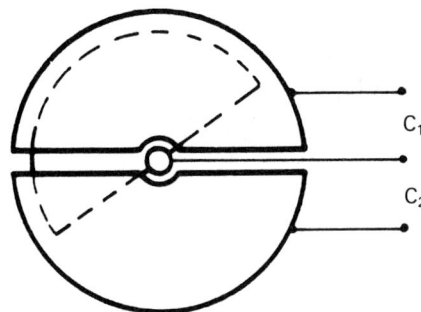

Figure 3.8   Differential variable capacitor.

The advantage of capacitors over resistive or inductive transducers is their low temperature dependency and the great accuracy which can be achieved. An example of a rotary capacitor used for high-precision measurement and computing tasks can be seen in Figure 3.9 which shows two mechanically coupled differential capacitors [13].

Each capacitor consists of a two-part or four-part stator and a rotor. As it is a differential capacitor, the capacitance between the rotor and the complete stator remains constant irrespective of the position of the rotor.

Different rotor shapes, Figure 3.10, can be used to produce different mathematical and empirical functions of the capacitor as a function of the angle of rotation $a$, e.g. linear, quadratic and trigonometric functions (sin, cos, $a/2$), Figure 3.11.

What must be emphasized are the high degrees of accuracy which can be achieved using such capacitors. For example, in the highest

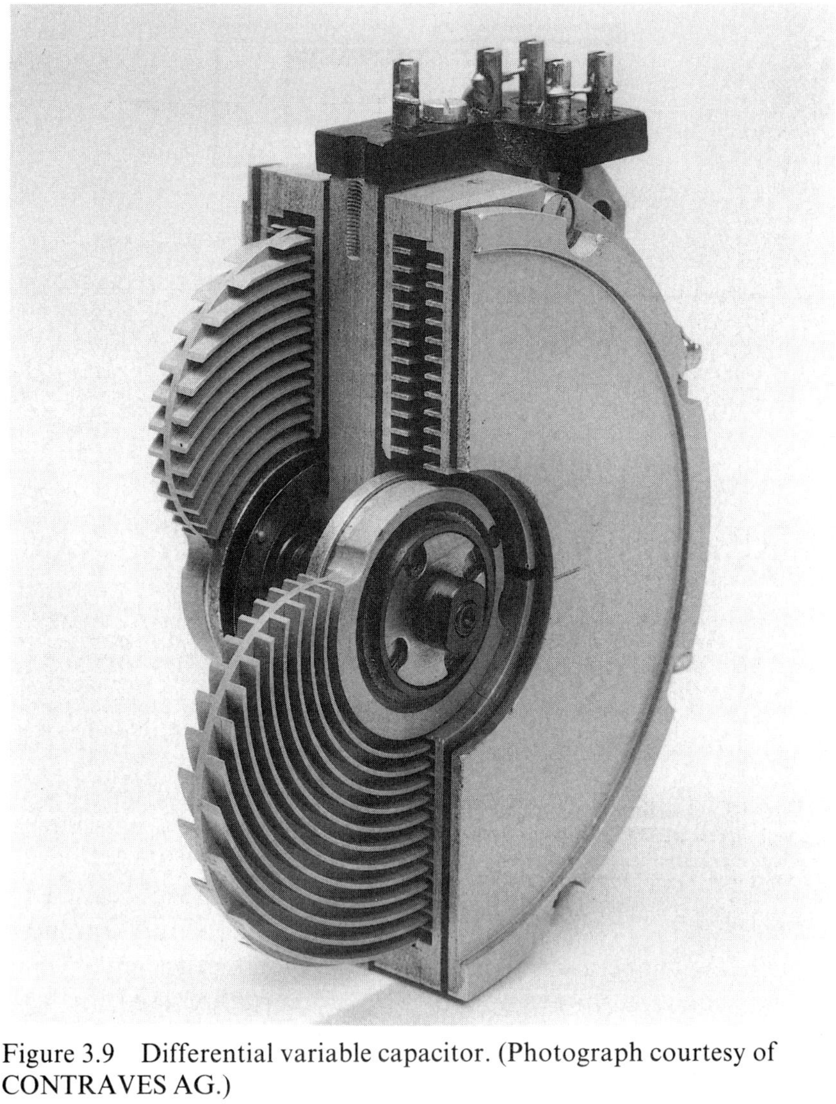

Figure 3.9 Differential variable capacitor. (Photograph courtesy of CONTRAVES AG.)

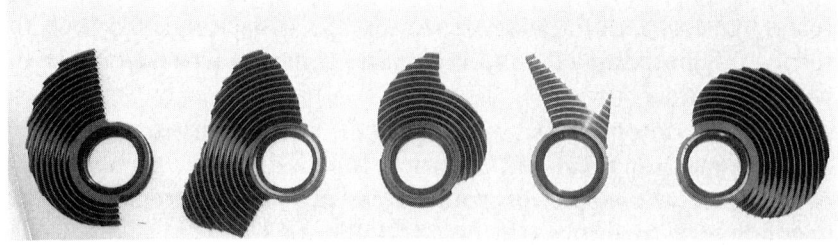

Figure 3.10 Rotor shapes for producing different functions for the capacitor. (Picture courtesy of CONTRAVES AG.)

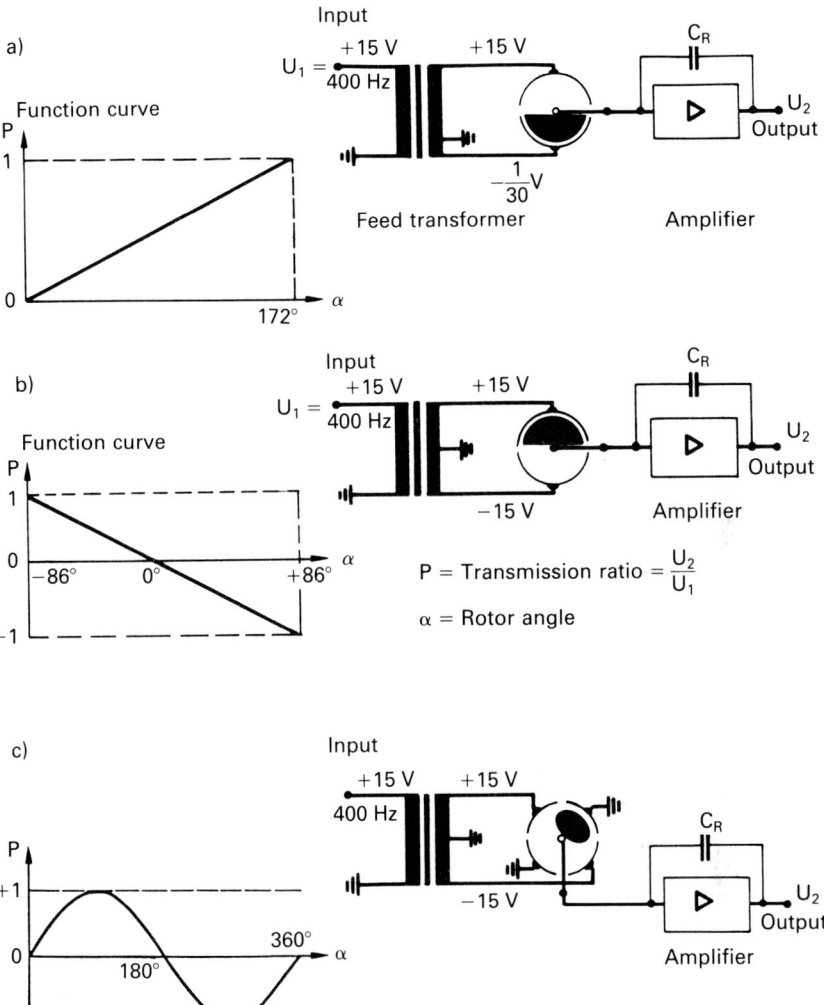

Figure 3.11 Function course and circuitry of differential capacitors. (a) Linear function of a capacitor with asymmetrical power supply. Working range 172°. (b) Linear function of a capacitor with symmetrical power supply. Working range 172°. (c) Trigonometric function. Working range 360°.

accuracy class for linear or trigonometric function, an error of ±0.05% relative to the maximum value of the function is guaranteed. The temperature coefficient is between $30 \times 10^{-6}$ and $60 \times 10^{-6}$ per degree Celsius.

### 3.1.4   *Inductive transducers*

The inductance of a coil is calculated from the relation:

$$L = \frac{N^2 \mu_0 \mu_r A}{l}$$

Where $N$ is the number of turns, $\mu_0 = 4\pi \times 10^{-7}$ $H/m$ the permeability of empty space, $\mu_r$ is the relative permeability, $A$ the area and $l$ the length of the coil.

The quantities $N$, $\mu_r$, $A$ and $l$ can be altered for displacement measurement. The number of turns can be varied by means of a mechanical pick-up (slide), Figure 3.12a. The geometrical dimensions of an individual coil are, however, difficult to alter.

If, on the other hand, a coil is divided into two halves and the space between them is varied, Figure 3.12b, the magnetic coupling is altered and with it the resulting total inductance. Moving a ferromagnetic core (longitudinal armature) within the coil alters the permeability ($\mu_r$). In one of the conductive plates (transverse armature) brought close to the coil or in a conductive sleeve placed over the coil, eddy currents are induced whose field interacts with the magnetic field of the coil, thereby altering its inductance (Figure 3.12c, d).

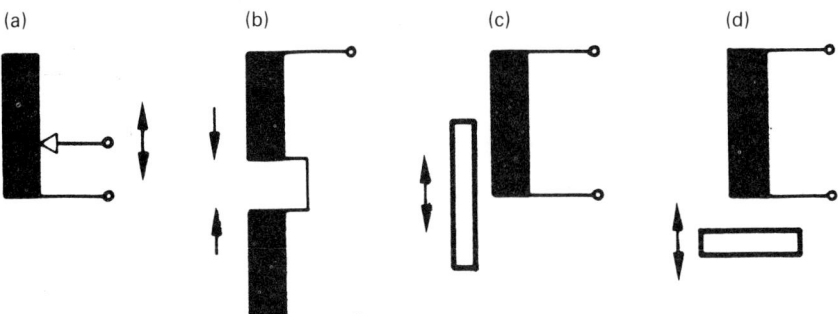

(a)         (b)         (c)         (d)

Figure 3.12   Altering inductance $L$ using: (a) $N$ number of turns, (b) geometric arrangement, (c) permeability (ferromagnetic core), (d) eddy current loss (conductive plate).

It is possible to measure displacement $s$ ranging from $1 \times 10^{-4}$ up to over 1000 mm using inductive transducers. Uncertainties of measurement less than 0.1% can be achieved. The most common con-

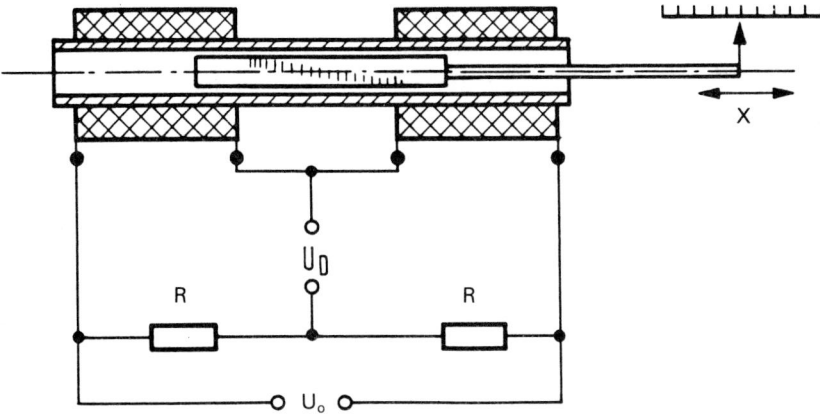

Figure 3.13    Inductive transducer in measuring bridge.

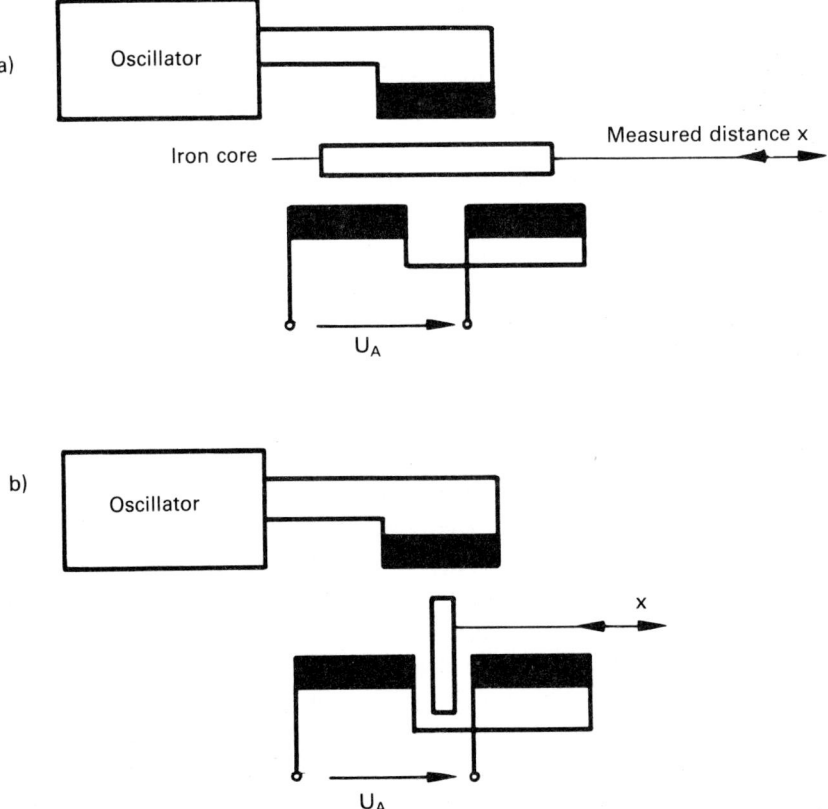

Figure 3.14    Principle of the differential transformer: (a) with longitudinal anchor, (b) with cross anchor.

structures are of half bridges (differential chokes) consisting of two coils and differential transformers with one primary and two secondary windings, Figures 3.13 and 3.14. The two basic types can be constructed with longitudinal or transverse armature.

Figure 3.13 shows a differential choke with longitudinal armature (ferromagnetic core). If the core is moved, the coupling is altered and with it the mutual inductance. The inductances in the two limbs of the bridge change in opposite directions and the bridge diagonal voltage shows the course displayed as a function of the measuring path $x$, Figure 3.15. The area around the zero crossing can be regarded as linear, to a very good approximation. Basically, two coils in differential connection are considerably less sensitive to stray magnetic fields, temperature fluctuations and fluctuations in amplitude or frequency of the supply voltage than single inductive resistors. Differential transformers are used even more frequently than differential chokes or half bridges with inductive transducers. The greatest degree of accuracy and sensitivity is achieved using longitudinal armature systems. The primary coil and the two secondary coils are wound round a cylindrical plastic coil former with a central bore. Depending on application, the plastic material must have high temperature stability and be very insensitive to humidity. This is why the windings are vacuum sealed in a plastic compound. In addition, the coil assembly can be mounted in a high-grade steel tube and sealed with a coat of epoxy resin. For particularly difficult environmental conditions, a high-grade steel tube is also placed in the inner bore and the tubes are welded to the front surfaces using high-grade steel plates, producing a hermetically sealed shell. Besides protecting against aggressive media, the casing provides a screen against magnetic and electrical fields. As with the differential choke, the displacement to be measured is converted into an electrical quantity by the movable iron core in the bore of the coil former. Power is usually supplied to the primary winding at a carrier frequency between 10 and 20 KHz. Depending on the position of the iron core, the magnetic coupling changes between primary and secondary winding. Two opposing alternating voltages of the same or different sizes are induced in both secondary coils. As the two secondary coils are connected in opposition to each other, the resulting output voltage is equal to the difference between the individual voltages. This is fed into a measuring amplifier. Figure 3.16 shows a control element for industrial use. An oscillator G generates the 10 kHz supply voltage for the primary coil LP of the differential transformer DWG82. The two secondary coils LS1 and LS2 are components of a measuring bridge the dia-

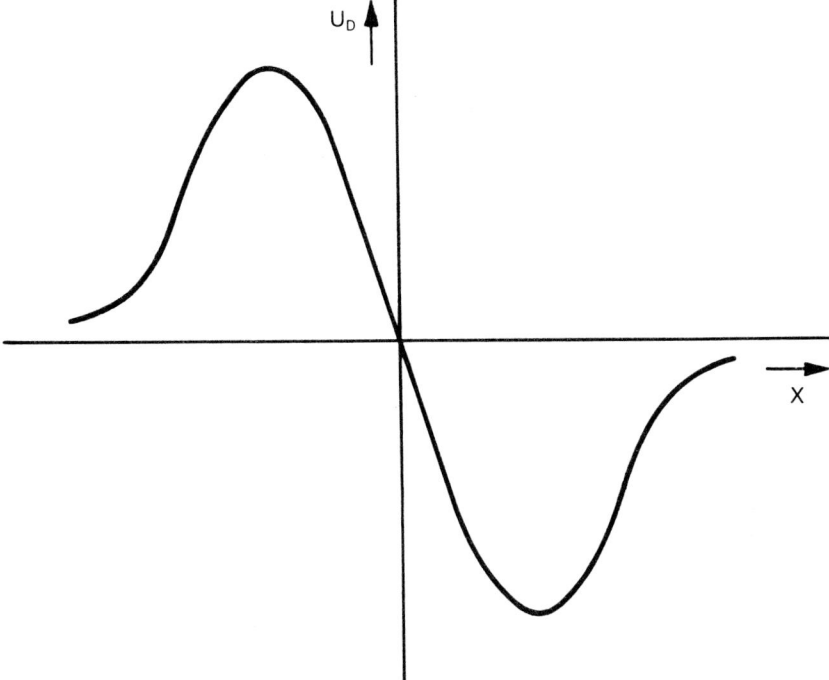

Figure 3.15   Bridge voltage with inductive transducer.

gonal voltage of which is fed via an impedance transformer to the measuring rectifier. A U/I transformer generates a standard signal 0(4) ... 20 mA.

This principle of measurement is also used for measuring angles (Figure 3.17). In this way the rotation is converted into a linear motion via a precision spindle (by moving the ferrite core in the differential transformer). All the electronics required to generate the 10 kHz supply voltage and to analyse the signal are housed in a sealed plastic casing [14].

The reaction power working on the iron core can be made so small as to be insignificant in the differential transformer if the core and the primary inductive resistor are properly designed. The characteristic curve of the output voltage as a function of the distance moved $x$ is similar to that of the differential choke. Coil assemblies both with and without an iron core are used for inductive displacement sensors. Although air-core coils have low inductances they can be used even at high frequencies. Iron cores produce greater changes in inductance and therefore higher sensitivities. Inductances with an iron core are,

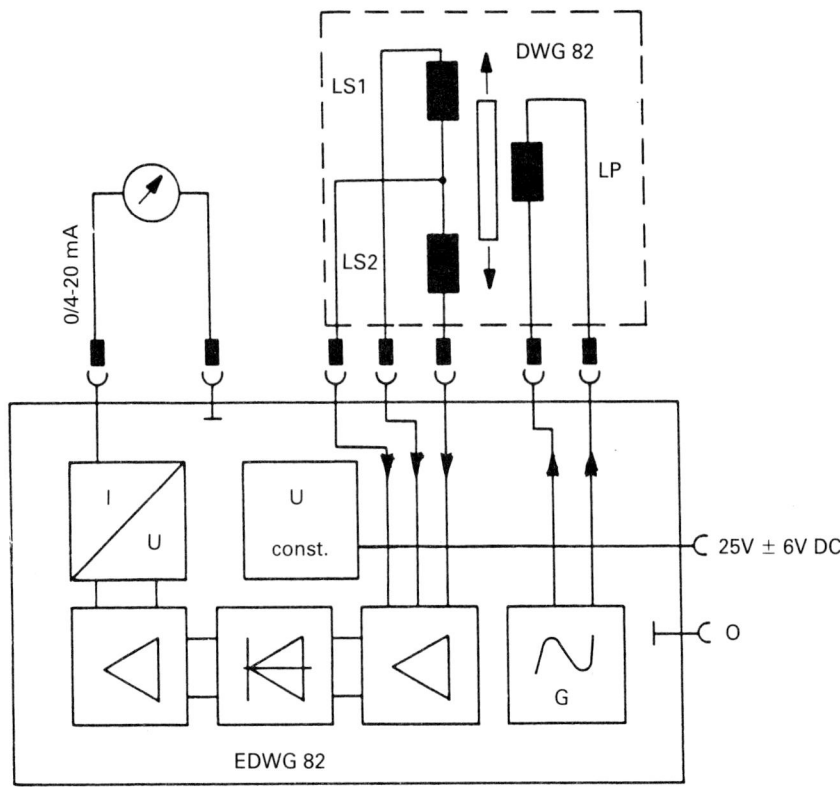

Figure 3.16   Differential transformer with circuit for generating output voltage and recognizing signals.

however, current-dependent and are subject to frequency-dependent losses [15;16]

Differential transformers have a number of characteristics which are not present in other displacement sensors and which make them especially suited to specific applications. The measurement is carried out in frictionless conditions as there is no mechanical contact between the movable core and the coils. This means it can be used for measurements in which no friction can be tolerated despite the light weight of the core, such as in vibration tests. The absence of friction has the further advantage of an unlimited mechanical lifespan as no wear is produced. This type of sensor is therefore suitable for applications requiring a high degree of reliability. The separation of coil and core enables the coil assembly to be isolated from pressurized or chemically aggressive media by means of a non-magnetic screen

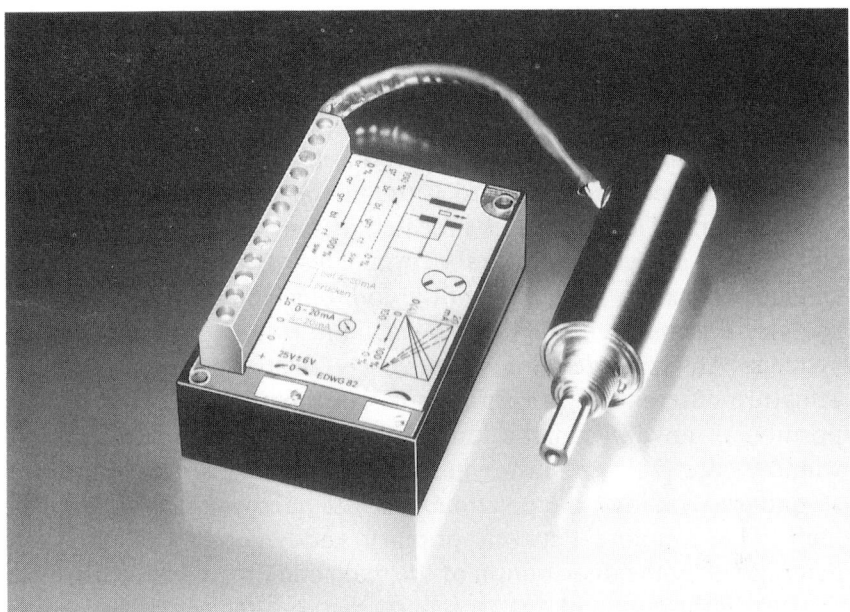

Figure 3.17    Angle measuring system: sensor (right) with supply and
recognition circuit (left). (Picture courtesy of Erwin Halstrup Multur
GmbH.)

between the inside of the coil assembly and the core. If the coils are
hermetically sealed, for example, if welded into a high-grade steel
housing, it is no longer necessary to provide the moving element with
a movable seal. Applications include rotary flowmeters, densito-
meters and level meters, as well as probes in hydraulic and pneumatic
control systems. Other outstanding characteristics include a
practically infinite resolution and an extremely high zero-point
reproducibility. For this reason, inductive sensors are very suitable as
null-balance indicators in control circuits. The electrical isolation of
the input from the output coil is valuable in applications where a
galvanic separation of systems with large electrical potential differences is required. Finally, the low sensitivity to lateral force must be
mentioned. This is important in applications for accelerometers
where, set at 90° to each other, there are three systems to detect the
three Cartesian components of any given acceleration vector [17].
The following data clearly indicate the ranges of application of
differential transformers: permissible ambient temperatures −200 to
+600°C, shock (11 ms) up to $1000\,g$; vibrations up to $20\,g$ at 2 KHz;
pressures up to 70 bar, linearity error 0.1%. These data do not refer to

any particular sensor, they are limiting values for various types which are best suited for specific characteristics required in certain applications.

In the case of transverse armature displacement sensors, a disc made of conductive material is introduced into the magnetic field, which causes eddy currents to form in the disc, resulting in flow displacement. The disc can be located at the end of the axis of the coil, at right angles to it, or in the form of a short-circuit sleeve which can move coaxially to the axis of the coil. In general, the sensitivity of transverse armature systems is lower than in longitudinal armature systems. An example of a type closely related to the transverse armature differential transformers uses a short-circuit ring which is positioned movably on the long limbs of a soft, magnetic, closed, window core [18]. As shown in Figure 3.18, two primary and two secondary windings are positioned at the narrow sides of the core limb. The coupling between primary and secondary windings is determined by the distribution of the magnetic flux, which consists of the flux in the core and any stray flux. Without the short-circuit ring the right and left coil assembly are symmetrical. The voltage $U_1$ and

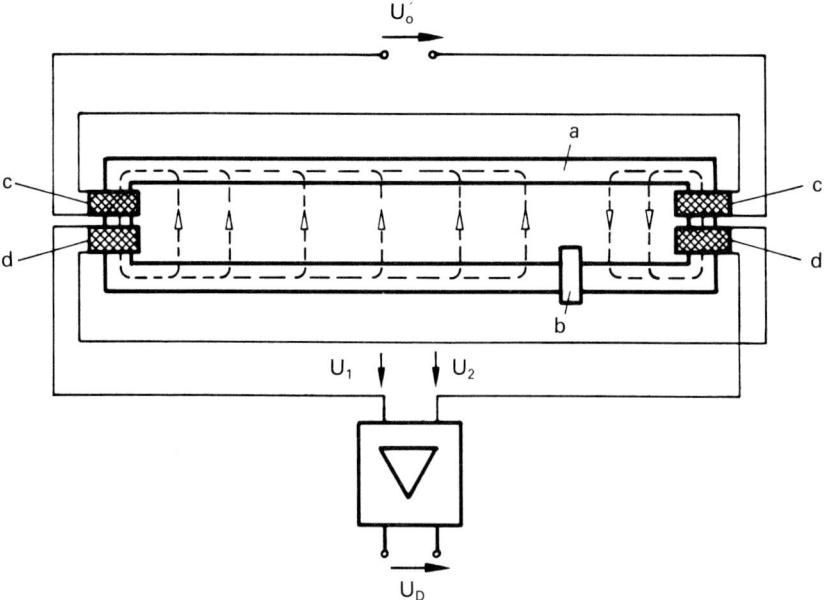

Figure 3.18   Differential transformer-type displacement sensor with short-circuit ring. a, Closed soft magnetic core; b, short-circuit ring; c, primary windings; d, secondary windings.

$U_2$ induced in both of the secondary coils are then equal and thus the differential voltage $U_D = 0$.

The short-circuit ring, which is positioned movably over one of the long limbs but not coming into contact with it, makes it possible to change the flux distribution in the window core and thereby the coupling between primary and secondary coils at a specified place. In ideal circumstances the short-circuit ring has the effect of an infinitely large, complex, magnetic resistor so that the magnetic fluxes close over the air-core to the right and left of the short-circuit ring. In other words, the ring acts like a virtual air-core. The course of the output voltages $U_1$ and $U_2$ is shown in Figure 3.19. The differential voltage $U_D$ has a maximum linearity error of 5% which can, however, be reduced to $\pm$ 0.5% by an electronic correction circuit [18]. Because of the very light weight of the short-circuit ring the measurement process is suitable for applications in which the object to be measured is subject to accelerations up to $300\,g$. Metal protection inside a sealed plastic housing enables it to be used at an ambient temperature of $180°$ in an extremely aggressive atmosphere or in corrosive liquids. The measuring distance is 280 mm.

Voltage

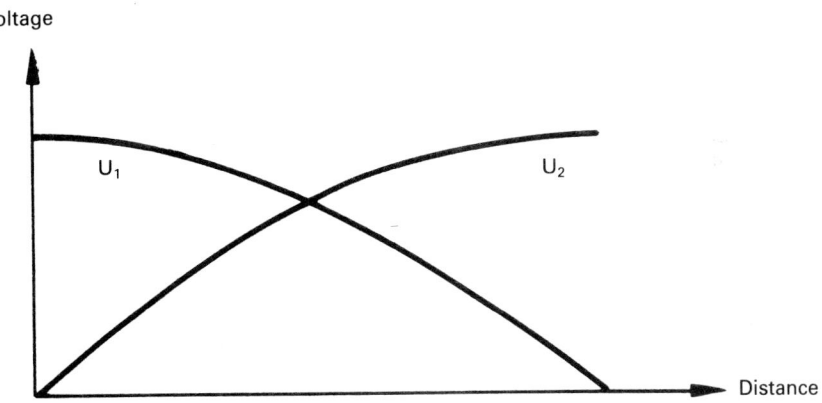

Figure 3.19   Voltage curve at secondary coils.

## 3.2   Features of processes using a digital–incremental measurement base

The term 'digital' comes from the Latin word digitus, meaning finger. This refers to elementary counting using the fingers. Counting, however, implies that the quantity to be counted has already been divided into equal parts or increments which can be readily counted. To achieve this digital–incremental measurement two procedures are required: quantization or division into increments, and counting. Quantization of the measurand occurs as soon as the material measures, e.g. a grid scale or a graticule, is produced. Counting is part of the measuring procedure. Incremental measurement processes are the simplest digital measurement processes using the principle of material measure, data acquisition and data processing. A further advantage is the feature whereby the zero point can be selected as desired by resetting or presetting the counter whereas a computation is required with absolute indicators. A previous serious disadvantage, that position data would be lost in the event of power failure, has been eliminated since the development of cheap non-volatile storage (EEPROM). Secure data transfer is being mastered nowadays, as the large number of incremental systems in use will testify.

A grid scale is composed of sections with different physical features, Figure 3.20, e.g. light and dark areas in optical systems. The distance between similar sections is the scale interval $T$. Counting the sections along a measured length A– B provides the position data. The material measure in the form of a grid scale or a graticule is a binary read-only memory which is interrogated as a function of the displacement or angle. One can think of many physical principles which are suitable for bringing about the two states of this memory and its reproducible interrogation. Of the numerous developments in this field, however, only a few have come to fruition, Table 3.1.

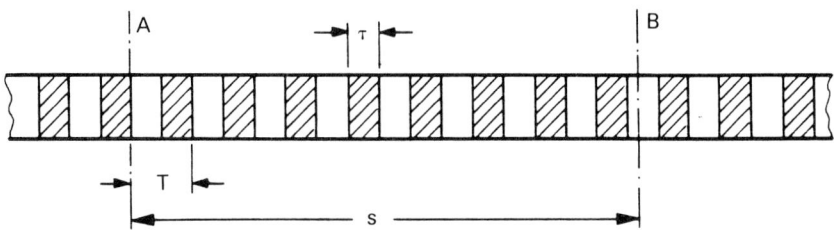

Figure 3.20   Digitizing a measured length. $s$, Measured length; $T$, grating period; $t$, increment.

A simple angle measurement process, Figure 3.21, consists of a cam wheel which makes an electrical contact on each rotation thereby moving forward by one increment the totalizer connected to

Table 3.1   Properties of the measurement base and associated sampling processes

| Measurement base | Sampling process |
|---|---|
| Variable conductivity | Mechanical sampling with contact<br>Springs or pins<br>Capacitative sampling<br>Sampling by frequency-limited element in<br>    oscillating circuit |
| Variable magnetism | Hall probe or ferrite core |
| Variable colour | Optical sampling by reflected light |
| Variable transparency | Optical sampling by transmitted light |
| Variable thickness | Measurement of phase difference between<br>    light beams |

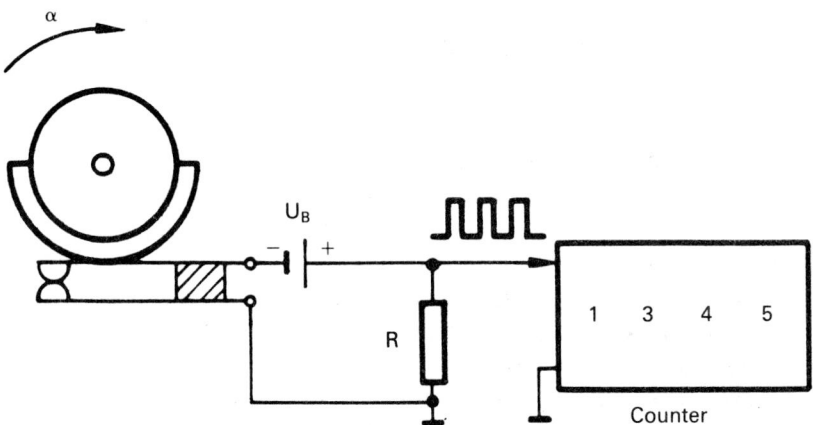

Figure 3.21   Simple displacement measuring system (one pulse per rotation).

it. So long as the direction of rotation does not change, an exact measurement can be taken using this type of apparatus, which generates a simple pulse train. Examples of applications for such measuring systems may include processing machines in which the product to be processed moves in one direction only and the longitudinal motion for measurement is converted to rotation (friction wheel). A preset totalizer can be used to count, starting from zero until the preset number is reached, then emit a signal or a procedure branching statement and finally return the totalizer automatically to zero and restart the procedure. One application would be the automated sawing of a piece of timber into sections of a specified length.

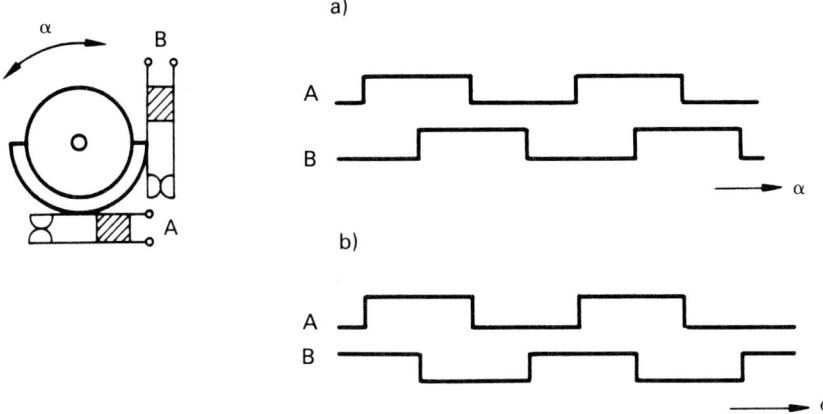

Figure 3.22   Generating two phase-shifted signals $A$ and $B$: (a) by anti-clockwise rotation; (b) by clockwise rotation.

The same can be achieved using a countdown counter which is set to a preselected value when work begins, stops a processing operation when zero is reached and is finally reset to the preselected value. If, however, the measurement is to be accurate even when the direction is reversed, then a second signal is required to recognize the reversal of direction. The two signals $A$ and $B$ must be separated by a phase shift. Depending on the direction of rotation, Figure 3.22, signal $A$ lag or lead signal $B$. This phase shift is usually selected at $T/4 + nT$), so that the resolution can be doubled or quadrupled in addition to directional recognition. $T$ is cycle duration of signals $A$ and $B$.

### 3.2.1   *Direction recognition and signal multiplication*

In order to recognize the direction of motion of an incremental trans-ducer it is necessary to check whether the $B$ signal is changing over from $O$ to $L$ or vice versa during the time when the $A$ signal is at level $L$. In the former case the direction of motion is forwards and in the latter case backwards, Figure 3.23. The evaluation of the edge of an impulse is effected by means of differentiation, which at its simplest is carried out by means of an RC module with a diode placed in parallel to suppress the negative impulses. In addition to the directional recognition, with the two signals $A$ and $B$ exactly in quadrature the resolution can be doubled in a simple manner by linking the two

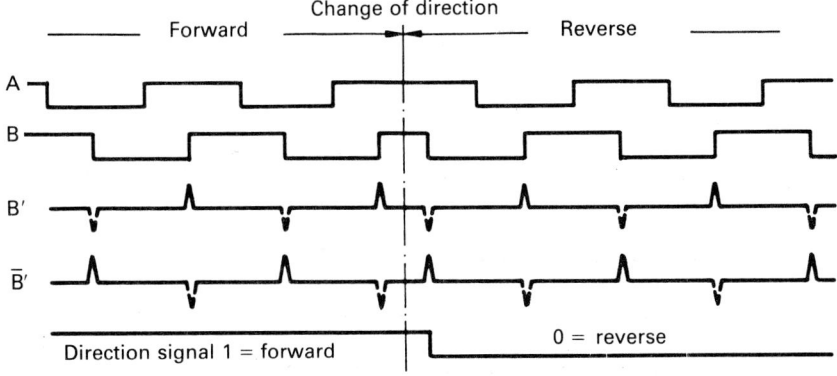

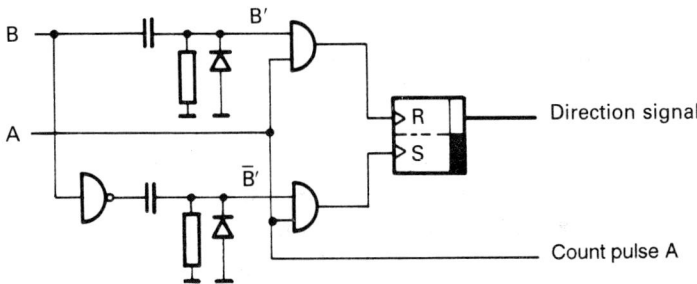

Figure 3.23 Direction sensing using incremental systems. *A, B*, Counting pulses; *B, B*, derived auxiliary signals.

signals via an exclusive OR gate. It is possible to quadruple the signal by means of the dynamic analysis of signals *A* and *B* and the exclusive OR signals *Ā* and *B̄*. In addition, the signals must be differentiated and the impulses originating at the positive edges must be combined via an OR gate, Figure 3.24. Integrated monostable circuits are frequently used instead of discrete RC modules with downstream pulse-forming stages. In addition to signal quadrupling, Figures 3.25 and 3.26 show the second method of providing an incremental display of position data. Unlike the circuit shown in Figure 3.23, in which a counting pulse and a direction signal are generated, in this case there are two pulse output points, one of which supplies a signal during forward motion and the other during backward

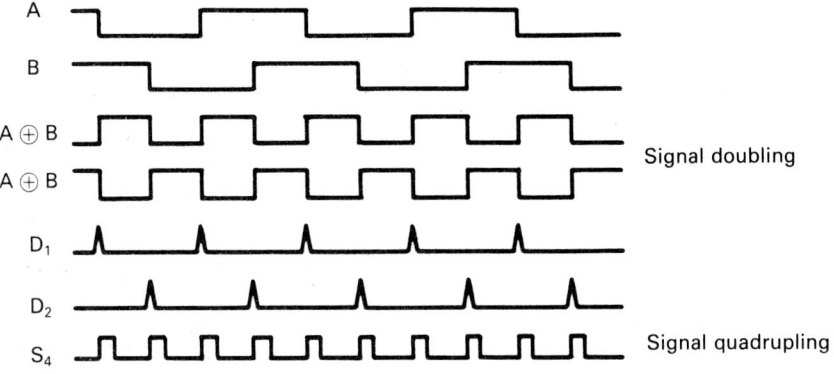

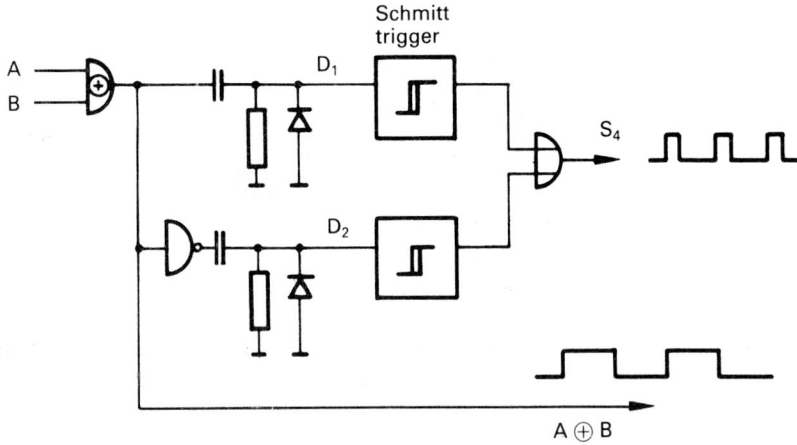

Figure 3.24    Circuit and clocking diagram for signal doubling and octupling.

motion. By using an OR gate to combine the metering pulses, a single counting pulse and, using a set–reset flip-flop as in the circuit in Figure 3.23, a directional signal can again be generated. By using $N$ scanners which are spaced at $a = T/2N$ relative to each other, an increase by a factor of $2N$ over the simple scanning of a grid can be achieved. It is possible to multiply a signal by eight by using four photodetectors, spaced one eighth of a scale division apart, to supply four signals out of phase by 45. The principle is shown in Figure 3.27,

Monoflops

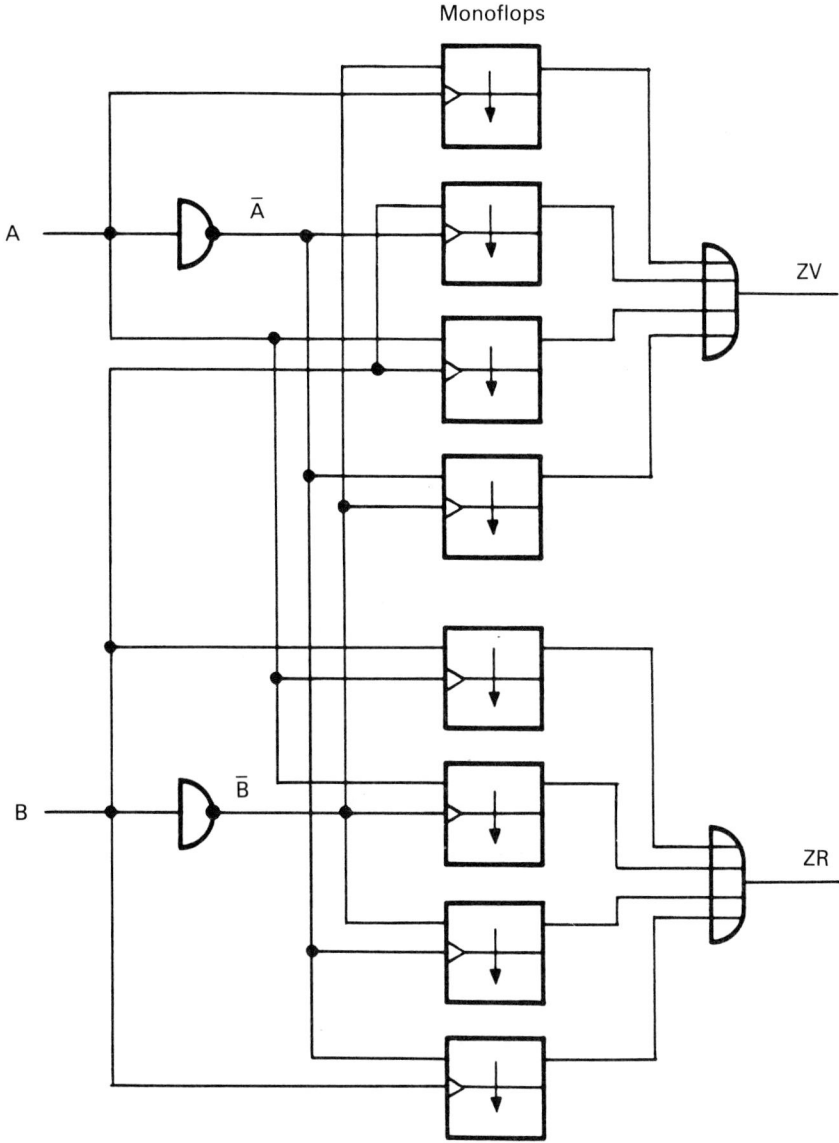

Figure 3.25 Signal quadrupling by means of monoflops.

where it should be noted that it is solely for the purposes of demonstrating the principle that the photodetectors are arranged in such a close sequence. If a spatial displacement of $a$ is required, then a displacement of $a + nT$ will also lead to a correct result, where $T$ is the interval of the scale division and $n$ is a whole number. In very

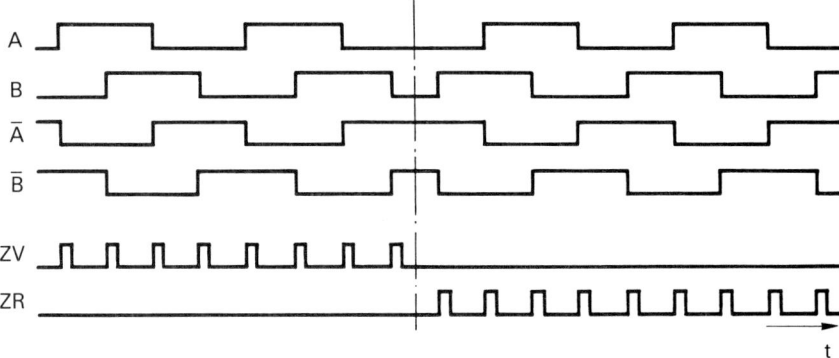

Figure 3.26   Generating count-up pulses $ZV$ and count-down pulses $ZR$ in signal quadrupling. A, B, Ā, B̄, Signals.

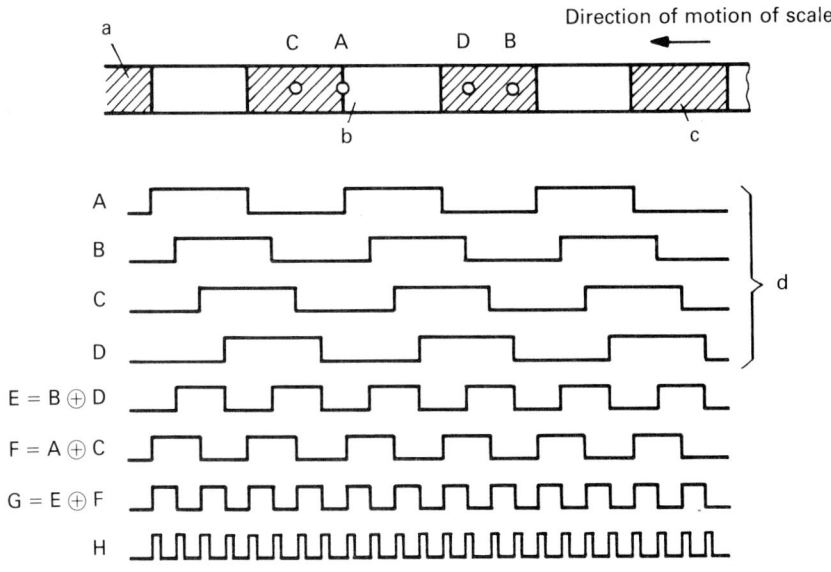

Figure 3.27   Arrangement of scanning points in signal octupling and respective clocking diagram. a, Scale; A, B, C, D, scanning points; b, transparent area; c, opaque area; d, photoelectric transducer signals; H, pulse train derived from the edges of G.

fine scale intervals $n$ can become very large. An increase in resolution by a factor of 16 assumes 8 scanners spaced 1/16th of a division apart.

Table 3.2   Assigning a number between 0 and 3 to input signals $A$ and $B$

| Condition | | Binary No. $Y_B$ | | Decimal No. $Y_D$ |
|:---:|:---:|:---:|:---:|:---:|
| A | B | $2^1$ | $2^0$ | $10^0$ |
| 0 | 0 | 0 | 0 | 0 |
| 1 | 0 | 0 | L | 1 |
| 1 | 1 | L | 0 | 2 |
| 0 | 1 | L | L | 3 |

## 3.2.1.1   Direction recognition by an adding device

An interesting possibility for direction recognition during simultaneous signal quadrupling allocates to each of the four possible states of signals $A$ and $B$ a binary number $Y_B$ [19]: in Table 3.2 each of the four possible states of input signals $A$ and $B$ is allocated a number between 0 and 3. Here $2^0 = A\ B$ (exclusive OR) and $2^1 = B$. The direction of rotation can be deduced from the direction the sequence of numbers pass through. Rising numbers 0, 1, 2, 3, 0 ... denote forward motion, descending numbers 0, 3, 2, 1, 0 ... reverse motion.

If two consecutive numbers are now subtracted $Y_n - Y_{n-1}$, the result is $-1$ for forward motion and $+1$ for reverse motion, ignoring

Table 3.3   $(Y_n + \bar{Y}_{n-1})$ with forward motion

| $Y_n$ | | $\bar{Y}_{n-1}$ | | $Y_n + \bar{Y}_{n-1}$ | | |
|:---:|:---:|:---:|:---:|:---:|:---:|:---:|
| $2^1$ | $2^0$ | $2^1$ | $2^0$ | $2^2$ | $2^1$ | $2^0$ |
| 0 | 0 | 0 | 0 | 0 | 0 | 0 |
| 0 | L | L | L | L | 0 | 0 |
| L | 0 | L | 0 | L | 0 | 0 |
| L | L | 0 | L | L | 0 | 0 |

Table 3.4   $(Y_n + \bar{Y}_{n-1})$ with reverse motion

| $Y_n$ | | $\bar{Y}_{n-1}$ | | $Y_n + \bar{Y}_{n-1}$ | | |
|:---:|:---:|:---:|:---:|:---:|:---:|:---:|
| $2^1$ | $2^0$ | $2^1$ | $2^0$ | $2^2$ | $2^1$ | $2^0$ |
| 0 | 0 | L | 0 | 0 | L | 0 |
| 0 | L | 0 | L | 0 | L | 0 |
| L | 0 | 0 | 0 | 0 | L | 0 |
| L | L | L | L | L | L | 0 |

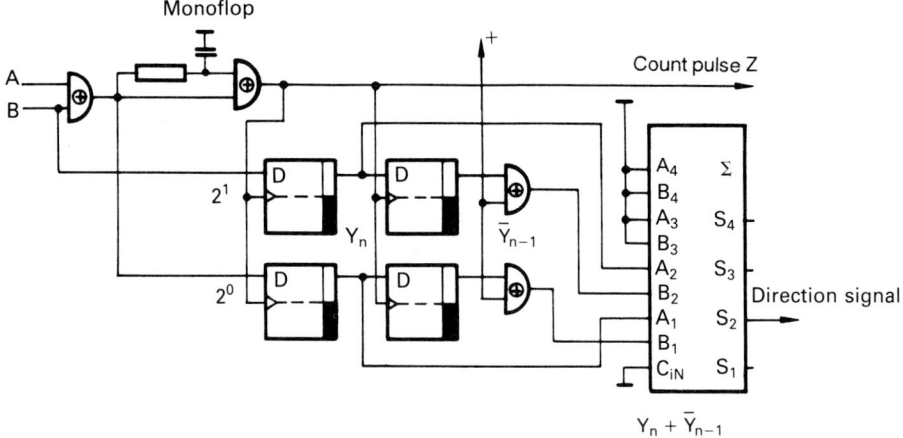

Figure 3.28   Circuit for quadruple analysis and direction sensing using a 4-bit adder.

any overflow. The subtraction is taken back to the addition of the complements of the subtrahend, Table 3.3. $Y_n - Y_{n-1} \rightarrow Y_n + \bar{Y}_{n-1}$ (the addition of 1 to the result, as required in exact subtraction, is not necessary).

If the sums arrived at are now compared, the digit $2^1$ is always $O$ for forward motion and always $L$, Table 3.4, for reverse motion, thereby becoming a criterion for the direction of motion.

In the circuit, Figure 3.28, signal $2^1$ is generated from signal $B$ and signal $2^0$ is generated via an exclusive OR gate. Signal $2^0$ therefore displays a doubling of the pulse at the same time. By evaluating the edges of $2^0$ with an exclusive OR gate as a monostable circuit, a metering pulse $Z$ of a quad frequency is generated for input signal $A$. A first pair of D flip-flops is used for synchronization with the counting pulse. After a delay for a further pulse, the complement of signals $Y_{n-1}$ is formed by two exclusive OR circuits. All four signals are fed to a 4 bit full adder whose output $S_2$, which has the significance $2^1$ (the total), denotes the direction. In the circuit indicated, attention should be paid to signal running times and to whether the time constant of the RC module (a few microseconds) is sufficiently small. A false directional signal may occur after start-up by virtue of the initially undefined states of the D flip-flops for one pulse. In order to avoid this disadvantage, external clock-actuated circuits are used in directional recognition and signal multiplication. In addition, these circuits have the advantage that they do not require RC modules to differentiate the edges and are therefore better suited for integration into solid-state circuits.

## 3.2.1.2   *External clock-actuated direction recognition*

Direction recognition by direct evaluation of the signal edges of two out-of-phase signals $A$ and $B$ may result in errors of measurement if, due to mechanical vibration of the incremental transducer, very short pulses are generated in which, at the same time, the direction of counting alternates very quickly. The superposition of input signals with interference pulses can also lead to counting errors unless these interference pulses are adequately suppressed by preceding low-pass filters. A considerably increased immunity to interference can be achieved by the use of external clock pulse-controlled circuits [20;21]. The basic idea of this is not to analyse the signal edges themselves but the bit pattern (or code) as determined, at specific points in time which have been determined by the external clock pulse, by signals $A$ and $B$ and two signals dephased by a specified time for this purpose. The process is therefore similar to the recognition circuit using the 4 bit full adder described in the previous section. Unlike the previous example, however, in which the control pulse was obtained from the signals of the incremental transducer, in this case an external pulse is used. Figure 3.29 shows the basic circuit in which signals $A_0$ and $B_0$, synchronized with the external pulse $T_F$, and signals $A_1$ and $B_1$, each dephased by a cycle of $T_F$, are obtained from $A$ and $B$ by means of two D flip-flops. The operation of a D flip-flop is characterized by any change in the data present at the preparatory input terminal D (e.g. signal $A$) not immediately appearing at the output but at the next positive clock pulse edge of $T_F$. This means that an input signal which is shorter than $t_1$ does not appear at the output if it is in the gap between two positive clock pulse edges of $T$. For this reason, only signals with the minimum duration $t_1$ of the clock pulse period occur at the output of the D flip-flop, which protects downstream recognition and counting circuits from malfunctions due to excessively high signal frequencies. When counting pulses are generated, the possible signal states of $A_0$, $B_0$, $A_1$ and $B_1$ are considered, Figure 3.30. Eight states (codewords) are possible for each of the forward direction and reverse direction, Table 3.5. This means that an analysis of the edges of input signals $A$ and $B$ is replaced by a comparison of a given signal with its associated time-delayed signal. A difference between $A_0$ and $A_1$ or $B_0$ and $B_1$ is the criterion for stating

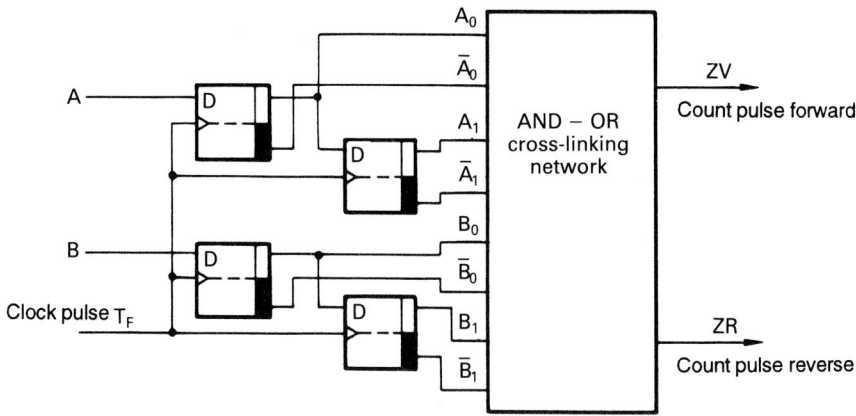

Figure 3.29   Externally timed signal quadrupling and direction sensing.

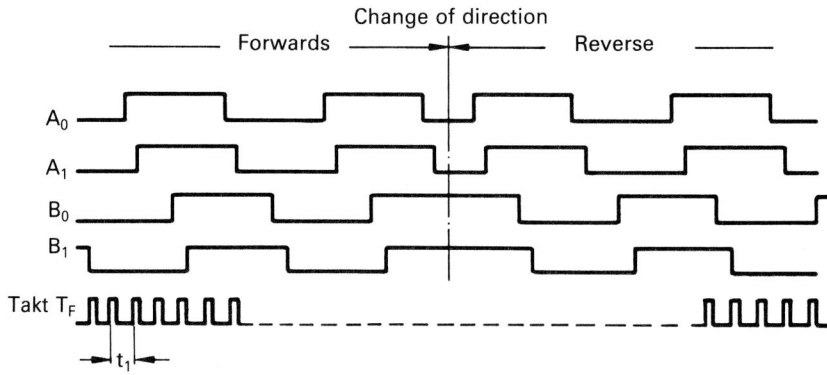

Figure 3.30   Clocking diagram for externally-timed signal analysis.

Table 3.5   Signal condition for forwards and reverse motion

| Condition / Signal | forwards | | | | | | | | reverse | | | | | | | |
|---|---|---|---|---|---|---|---|---|---|---|---|---|---|---|---|---|
| | 0 | 1 | 2 | 3 | 4 | 5 | 6 | 7 | 8 | 9 | 10 | 11 | 12 | 13 | 14 | 15 |
| $A_0$ | 0 | 1 | 1 | 1 | 1 | 0 | 0 | 0 | 0 | 0 | 0 | 1 | 1 | 1 | 1 | 0 |
| $A_1$ | 0 | 0 | 1 | 1 | 1 | 1 | 0 | 0 | 0 | 0 | 0 | 0 | 1 | 1 | 1 | 1 |
| $B_0$ | 0 | 0 | 0 | 1 | 1 | 1 | 1 | 0 | 0 | 1 | 1 | 1 | 1 | 0 | 0 | 0 |
| $B_1$ | 0 | 0 | 0 | 0 | 1 | 1 | 1 | 1 | 0 | 0 | 1 | 1 | 1 | 1 | 0 | 0 |

that an exchange of signals has taken place. Only when a signal transmission from $O$ to $L$ or vice versa has taken place does a differential
signal occur. In order to ensure that all the edges of the transmitting
signal are analysed, the clock pulse frequency must be $f_T\, 8\, N$, where

$N$ is the number of signal cycles per second. In the case of shaft encoders this corresponds to the scale number on the disc multiplied by the number of revolutions per second.

When comparing the codewords which occur it will be noted that states 0, 2, 4, 6 (forwards) and 8, 10, 12 and 14 (reverse) are in matching pairs. For example, state 0 is identical to state 8, etc., and therefore not a criterion for a direction signal. Clearly, states 1, 3, 5 and 7 are for the forward direction and 9, 11, 13 and 15 for the reverse direction. Depending on the type of signal analysis (simple analysis, signal doubling or signal quadrupling) only one or all sixteen possible states are analysed. In simple analysis only the positive edges of $A$ are considered, i.e. the states in which signal $A_0 = 1$ and similarly $A_1$ show a value of zero. The expression for the counting pulses is therefore:

in the forward direction

$$ZV = A_0 . \bar{A}_1 . \bar{B}_0 . \bar{B}_1 \tag{29}$$

in the reverse direction

$$ZR = A_0 . \bar{A}_1 . B_0 . B_1 \tag{30}$$

In the case of double analysis the negative edge of $A$ is also considered:

forward direction

$$ZV = A_0 . \bar{A}_1 . \bar{B}_0 . \bar{B}_1 + \bar{A}_0 . A_1 . B_0 . B_1 \tag{31}$$

reverse direction

$$A_0 . \bar{A}_0 . B_0 . B_1 + \bar{A}_0 . \mathbf{A}_1 . \bar{B}_0 . \bar{B}_1 \tag{32}$$

When all the edges are analysed, a signal quadrupling is obtained: counting pulses forwards

$$\begin{aligned} ZV = &\bar{A}_0 . \bar{A}_1 . \bar{B}_0 . \bar{B}_1 + \\ &\bar{A}_0 . A_1 . B_0 . \bar{B}_1 + \\ &\bar{A}_0 . A_1 . B_0 . B_1 + \\ &\bar{A}_0 . \bar{A}_1 . \bar{B}_0 . B_1 \end{aligned} \tag{33}$$

counting pulses in reverse

$$\begin{aligned} ZR = &\bar{A}_0 . \bar{A}_1 . B_0 . \bar{B}_1 + \\ &A_0 . \bar{A}_1 . B_0 . B_1 + \\ &A_0 . A_1 . \bar{B}_0 . B_1 + \\ &\bar{A}_0 . \bar{A}_1 . \bar{B}_0 . \bar{B}_1 \end{aligned}$$

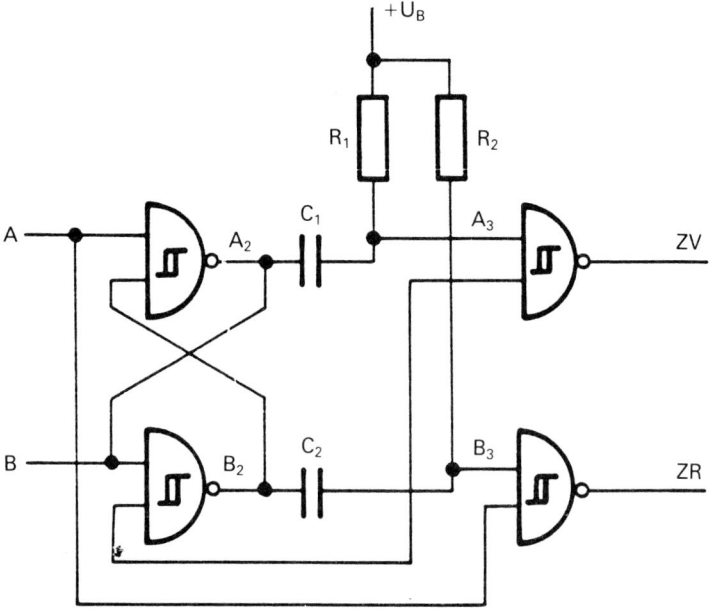

Figure 3.31    Simple circuit for generating count-up and count-down pulses.

A very simple and economical circuit for direction and its associated timing diagram are shown in Figure 3.31. This supplies one counting pulse $ZV$ per signal cycle for the forward direction and $ZR$ for the reverse direction. At the same time it suppresses interference due to mechanical vibration, provided these are smaller than one signal cycle. The width of the output pulses is determined by the time constants $R_1 C_1$ or $R_2 C_2$ as the case may be. If $R_1 = 100\,\text{k}$ and $C_1 = 100\,\text{pF}$ are selected an output pulse $10\,\mu s$ wide is obtained [22].

A further circuit for direction recognition, which at the same time generates a signal quadrupling as well as reducing the effect of vibrations, is shown in Figure 3.32. Again, forward counting pulses $ZV$ and reverse-counting pulses, $ZR$ are obtained as output signals. The Schmitt triggers used as input stages increase immunity to interference by suppressing disturbance pulses on the encoder lines. A further two pairs of Schmitt triggers, each with an RC module, form two delay circuits each with a propagation delay time of $10\,\mu s$. In this way, a bit pattern as shown in Table 3.5, used as an entry point address for a 4 bit binary decoder, is produced from the two input signals $A$ and $B$ and the associated delayed signals. Each of the 8

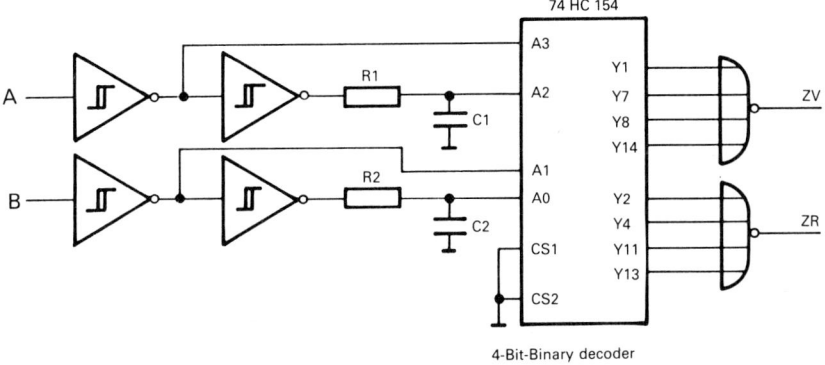

Figure 3.32 Using a 4 bit binary decoder to generate direction signals.

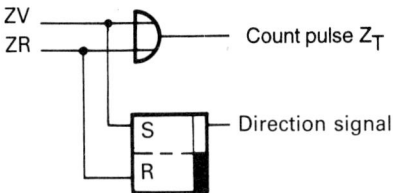

Figure 3.33 Producing the counting cycle and direction signal from count-up and count-down pulses.

possible combinations in all (forward and back movement) of $A$ and $B$ in which each of the edges is viewed results in a corresponding output signal $ZV$ or $ZR$ [23].

A common counting cycle and a direction signal can be generated, if required, from the separate counting pulses for the two directions of movement, Figure 3.33. To sum up, the decisive advantage of having movement direction discriminators synchronized by external timing cannot be emphasized enough: interference pulses on the input lines either do not arise at all, if they fall in the space between two clock pulses, or they generate an output pulse at each of the two outputs $ZV$ and $ZR$ one after the other, if the disturbance pulse coincides with the clock pulse. A subsequent counter would therefore count once up and once down thereby ensuring no permanent miscounting would result.

## 3.2.2   *Signal multiplication with sinusoidal output signals*

Many high-resolution photo-electric systems generate sinusoidal out-put signals. These signals resemble resolver signals, which is why people refer to them as optical resolvers [24]. Unlike resolvers in which the modulation of a carrier frequency always takes place, in this case there are static signals when the system is stationary. This means that the sinusoidal curve is a function of distance and not time. Signal multiplication therefore calls for methods other than those used to digitize resolver signals. A very simple method suitable for lower resolutions makes use of the fact that by addition or subtraction of two sinusoidal signals $a < \sin x$ and $b < \cos x$ which have the same frequency but are phase-shifted by 90°, another sinusoidal signal $s$ is generated with a phase-shift dependent on the amplitude ratio $b : a$:

$$s = a \sin x \pm b \cos x = \sqrt{a^2 + b^2} \sin(x \pm \varphi) \tag{35}$$

where

$$\varphi = \arctan \frac{b}{a} \tag{36}$$

Consequently, further sine functions (theoretically any number) can be generated from one sine and one cosine function. The zero cross-ings of these functions supply the switching points for downstream trigger levels. Figure 3.34 shows two sinusoidal oscillations lagging or leading $S_1$ and $S_2$ by 45° being generated, where factors $a$ and $b$ are selected in relation to $l$. Trigger levels are used to generate four rectangular signals phase-shifted by 45°, from which a quadruple frequency signal can then be generated. This can again be doubled using two monostable flip-flops, thereby reaching a factor of 8. The basic circuit required to generate two signals phase-shifted by $\pm d$ in relation to $S_1$ is shown in Figure 3.35. Limitations are placed on the process by the cost of operational amplifiers and potential dividers, by the accuracy of the curve from and the quality of the amplifiers. In practical operation it is used up to multiplication by a factor of 25. What distinguishes it from other processes is its high conversion rates.

Additional methods of interpolation are created when, besides using a full-wave rectifier to add and subtract the sinusoidal primary signals, thereby obtaining a value, further signals are generated using other zero crossings [25], Figure 3.36.

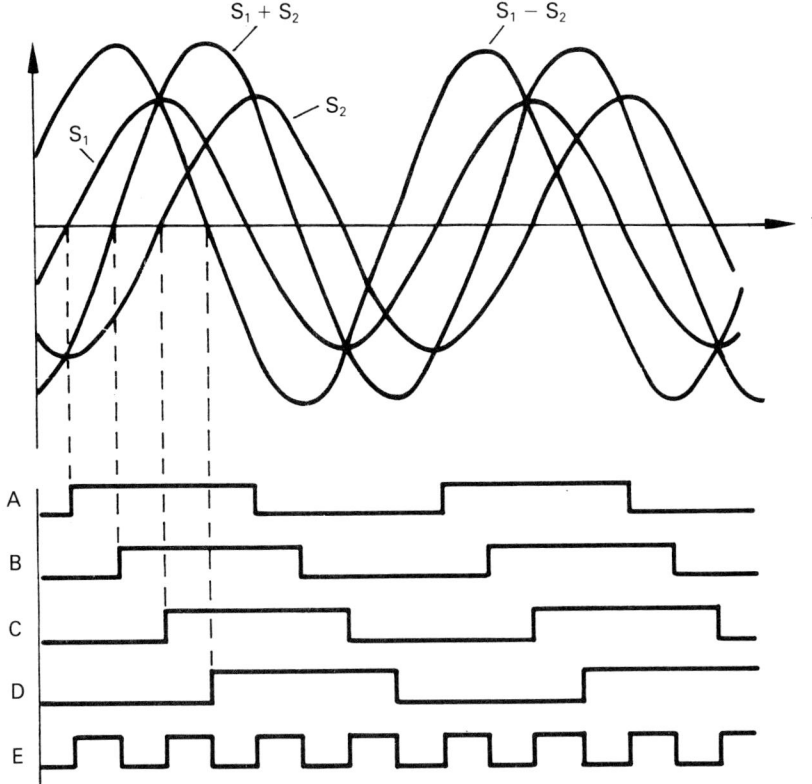

**Figure 3.34**  Generating four signals *A*, *B*, *C*, *D* in quadrature and one quadruple frequency signal *E* derived from them by adding and subtracting two sinusoidal signals $S_1$ and $S_2$.

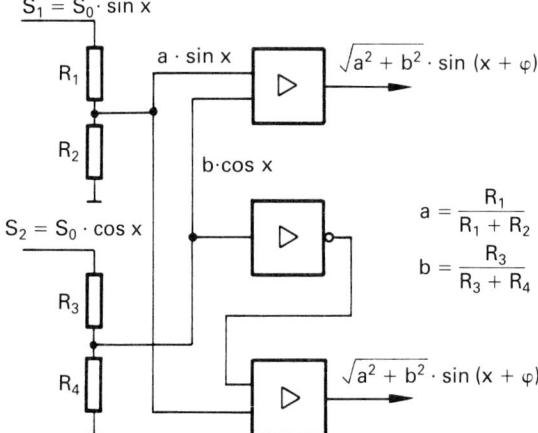

**Figure 3.35**  Circuit for generating phase-shifted sine signals.

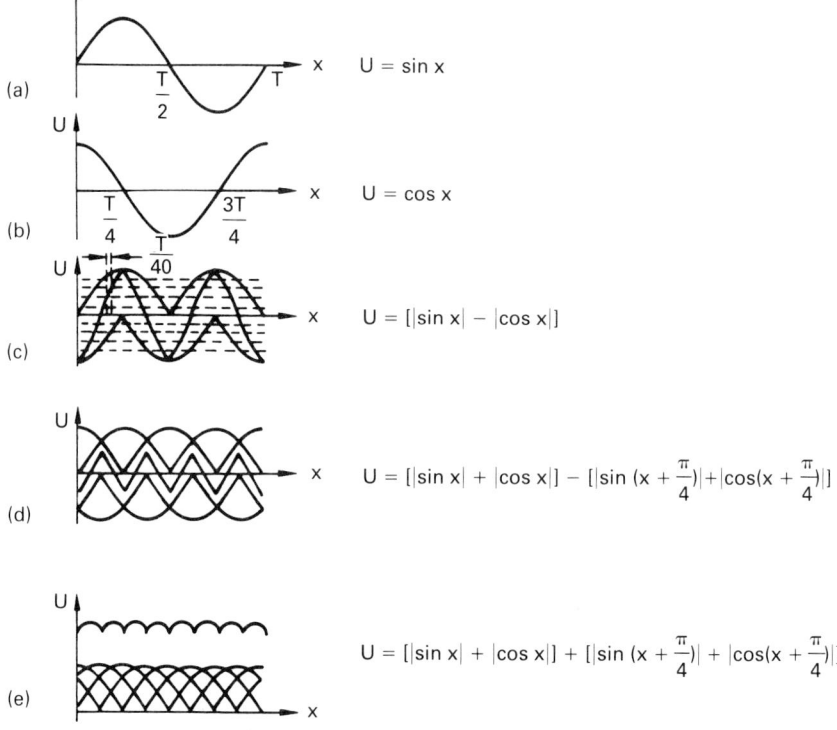

Figure 3.36   Generating additional zero crossings for signal multiplication by sixteen by rectification and differentiation or addition.

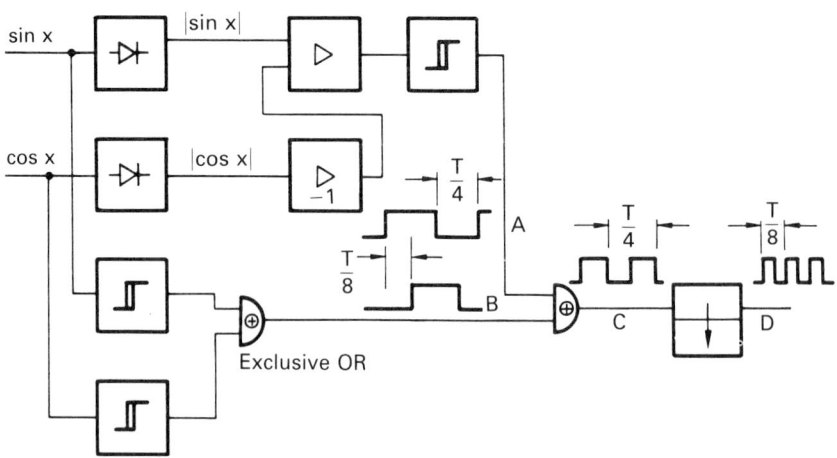

Figure 3.37   Circuit for signal octupling.

In order to achieve multiplication by a factor of 8, the input signals $S_1 = \sin x$ and $S_2 = \cos x$ are first triggered at their zero crossings and linked via an exclusive OR gate in the normal manner. This produces the usual signal duplication. A second signal phase shifted by 45° is obtained in which the difference $|\sin x| - |\cos x|$ is formed and triggers the approximately triangular signal which arises, Figure 3.36, again at the zero crossings. Two signals therefore arise, spaced at $T/4$ from each other, from which the desired signal is generated by means of a further exclusive OR gate and associated edge analysis, Figure 3.37.

It is also possible to multiply a signal by a factor of 16, based on the circuit for the octuple subdivision. An additional analogue signal is also required which has a cycle of $T/4$ and is out of phase by $T/16$ in relation to the signal $C$ (Figure 3.37) obtained from the octuple circuit. This analogue signal is also approximately triangular (Figure 3.36d). In order to generate it, two signals, each lagging the primary signals $\sin x$ and $\cos x$ by $\pi/4$, are obtained by addition or subtraction. These then have the curves $\sin(x + \pi/4)$ and $\cos(x + \pi/4)$. The desired course is produced by totalling and further additions and subtractions according to the following relation:

$$S_4^* = [|\sin x| + |\cos x|] - \left[\left|\sin\left(x + \frac{\pi}{4}\right)\right| + \left|\cos\left(x + \frac{\pi}{4}\right)\right|\right] \quad (37)$$

From $S_4^*$ a trigger level is used to generate the desired rectangular

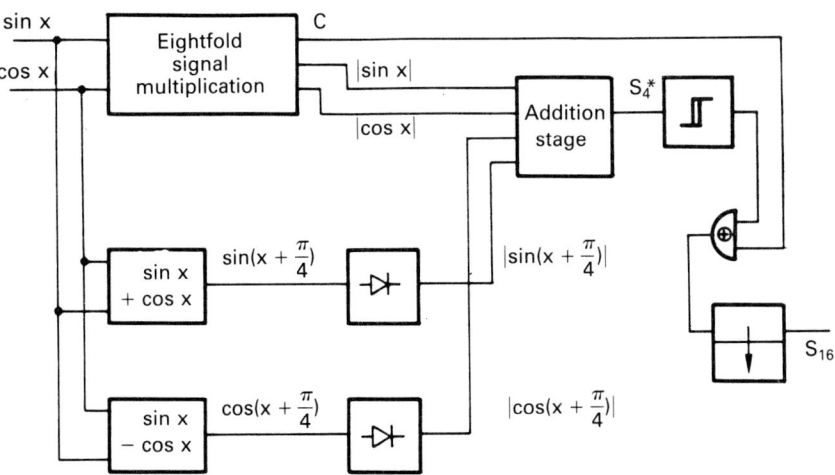

Figure 3.38 Multiplying signal sixteen times.

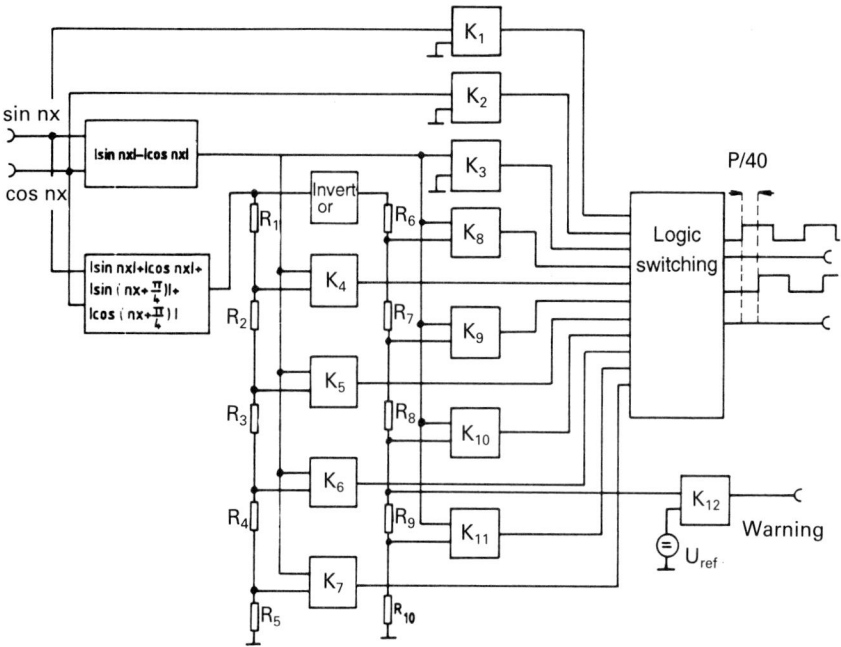

Figure 3.39   Multiplying signal forty times. (Picture courtesy of ZEISS.)

signal with a cycle $T/4$, phase-shifted by $T/16$ in relation to signal $C$. In this way, two signals phase-shifted by 90° are again obtained, but these already have the quad frequency of the primary signals. The desired 16-times frequency can then be obtained by the usual means (signal multiplication), Figure 3.38.

A further increase in the resolution up to a factor of 40 is possible if the signal as defined in equation (37) is fed to comparators with different switching thresholds. To this end, the reference inputs of the comparators must be located on voltage dividers with accurate and stable division ratios. In Figure 3.36 the various switching thresholds are plotted, which have been selected so that the scale spacing allocated corresponds to the interpolation interval $T/40$.

Figure 3.39 shows the block diagram for the 40-times operation. The comparators $K_1$ and $K_2$ are triggered by the primary signals, and the comparator $K_3$ by the near-triangular signal during the zero crossings. The near-triangular signal goes to an input of each of the comparators $K_4$–$K_{11}$, the reference voltages of which are determined by the voltage dividers $R_1$–$R_5$ for the positive half-wave and $R_6$–$R_{10}$

for the negative half-wave. It is not a constant direct-current voltage which is used as the input voltage for the voltage dividers, but a voltage as shown in Figure 3.36d. This reference signal, which is generated directly from the primary voltages, guarantees that the circuit will operate free from interference even in the event of changes to the amplitudes and phases of the primary signals. The rectangular signals originating at the outputs are linked in a gate network in such a way that two rectangular signals occur, the space between the edges of which are $T/20$ and the reciprocal phase shift of which is $T/40$. If 18 comparators are used instead of 8, then it is even possible to have 80 times the resolution [25].

A quite different approach to the interpolation of a signal phase is to calculate the angle $\varphi$ using the relation $\varphi = \arctan b/a$. It should be noted that for large values of $b/a$ where the tangent would be almost infinite, unacceptable errors may occur. It is therefore advisable to calculate $a/b$ and the cotangent, and not $b/a$, from an angle determined by system accuracy.

Another process uses a null-seeking circuit for digitization by means of a digital sine–cosine generator (see Figure 3.138). In this case a counter consisting of a sine–cosine table in read-only memory and digital–analogue converters is used to generate sine and cosine signals. These two signals are compared with the signals supplied by the incremental indicator and where counters deviate a forward or reverse count is carried out. The position of the counter when adjusted therefore shows the digital angle data.

The multiplications usual with optical resolvers go up to a factor of $2^8 = 256$ [24].

### 3.2.3   Reset pulse/reference signal

In the case of incremental systems, a reference signal is required to find a specific position again after power failure or shut-down. This reference signal is used to set the counter belonging to the measurement process to zero or another preselected value once the process has been restarted at a clearly specified position. The ability to preset the process to any desired value by means of reference signals is an advantage of the incremental method of measurement with its free choice of reference system.

As the reference or reset pulse in angle encoders occurs once per rotation, it can be used to store the counter position at that particular time in a non-volatile memory. If, in addition, the signal which indicated the direction of counting is stored, measurement can be continued after a power failure at the correct value as soon as the next reference signal occurs. It is not therefore necessary to start the machine at the zero point. This method is also used in incremental displacement measurement processes by placing a number of reference marks at fixed distances of, for example, 50 mm.

A further interesting possibility is to use the reference pulses to carry out an error check as there must be a specified number of individual pulses between two reference marks. (This does not, of course, apply where the direction is reversed.) Moreover, a separate count of rotation is possible.

Care must be taken with the position and width of the reference pulse to ensure that its edges do not coincide with those of counting pulses. Influences such as temperature, speed of rotation and component tolerances of the analysing electronics cause slight displacements between the counting pulses and the reference pulse. If, however, the front edge of the reference pulse appears once before and once after the appropriate counting pulse then the difference between the count in both cases is unity.

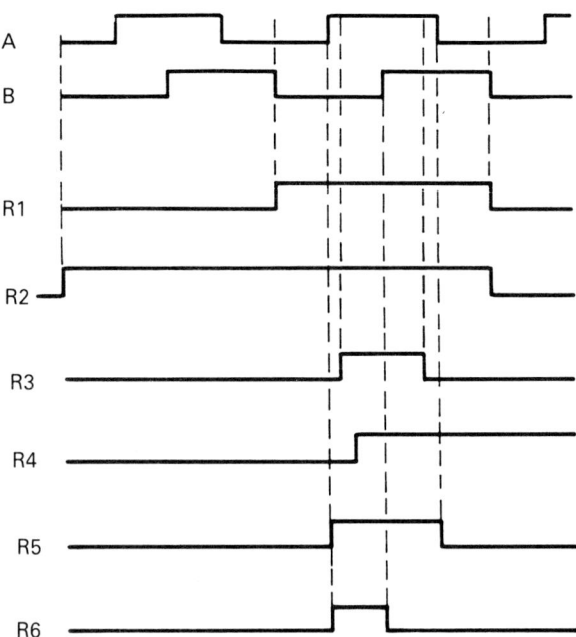

Figure 3.40  Common shapes of reference signal. *A, B*, Counting signals; *R*1...*R*6, reference signals.

The shapes of reference signal used in practice are shown in Figure 3.40. *R*1 is the reference signal most commonly used in practice. Its width is a lattice constant, which corresponds electrically to 360° and its edges are symmetrical relative to a pulse of the *A* signal. This arrangement means that the edges of the reference signal are in the middle of the permissible tolerance range. The reference signal *R*6, which can be generated by combining *A*, *B* and *R*1, is used particularly for higher pulse rates. In this way problems with the position tolerances of the edges can be more easily overcome.

In order to increase reliability when transmitting the reference signal, the relative inverted signal can be transmitted simultaneously. In the receiver, all common-mode interference is then rendered ineffective by subtraction. A different method is used by reference signal *R*2 which contains two consecutive phases of the *A* signal. The reference signal is said to be valid if and only if coincidence has resulted twice in succession at *A*.

A third possibility is reference signal *R*3, which lies completely inside a pulse of the *A* signal. Usually *T*/3, corresponding to 120° electrically is selected. Apart from the fact that the maximum permissible variations of pulse duration are considerably smaller than in the case of *R*1, greater difficulties are caused when generated at high pulse rates than in the case of *R*1, as, in optical systems, for example, very narrow slits which let little light through have to be used.

The fourth solution is to use a reference signal corresponding to *R*4. In angle encoders, a signal about an angle of rotation of 180° is mechanically generated with subsequent dynamic analysis of the edges. The disadvantage of this is that pulses dependent on the direction of rotation and of different polarity are obtained and the width of these pulses has no fixed relationship with the pulse width of signals *A* and *B*. The reference signal *R*5, which is equally common in practice, corresponds exactly in width to a pulse in track *A*. This apparent contradiction of the foregoing argument that allowance should be made for variations in environmental conditions and tolerance of components is only superficial as it has originated from a logical conjunction of *R*1 and *A*.

High-resolution incremental transducers today are mainly manufactured on the basis of photoelectric scanning of optical grids. Where gratings are very fine the generation of a reset pulse causes greater problems than the generation of the counting pulses, as one grating can be scanned by another whereas a zero mark must be scanned over a gap. The finer the grating becomes, the more the light available in gap-scanning diminishes. A very interesting solution to this problem consists of providing not a zero mark for the reset pulse, but a relatively wide field with a special grating. What is special about this grating is that the distance between the lines is not equal but has been specified in accordance with a sequence of random numbers. If such a pattern of lines is moved past a second grating, which is a copy of the first, then the number of points of agreement in relation to the bright field will increase slowly at first but will jump straight to a value covered by both gratings, which is clearly different from those preceding it. In mathematical terms the digital autocorrelation function of both pulse trains is being formed. The expression for the autocorrelation function is:

$$R_{xx}(\tau) = \lim_{T \to \infty} \frac{1}{2T} \int_{-T}^{+T} x(t)\, x(t + \tau)\, dt \tag{38}$$

This can be used to determine the statistical relationship between two signal sections:

$$x(t) \text{ and } x(t + \tau).$$

The autocorrelation function is the mean value of $x(t) \times x(t + \tau)$ and the mathematical operations are delay, multiplication and averaging.

In digital signals coincidence is the equivalent of multiplication and totalling equates to averaging. The autocorrelation function is then:

$$R_{xx}(\tau) = \frac{1}{N} \sum_{k=1}^{k=N} x(k\Delta t)\, x(k\Delta t + \tau) \tag{39}$$

As the signals $x\,(k\Delta t)$ and $x\,(k\Delta t + \tau)$ can only take on one of the two values 0 and 1, value 1 is given for their product only if they both have value 1.

This is equivalent to coincidence by means of conjunction.

Figure 3.41 shows a very simple example of a pseudo-pulse sequence (a) and the corresponding grating (b) required to generate a reset pulse with the width of a lattice constant $T$ of the ruled grating (c). If the pseudo-division is now used for the zero mark on the movable index grating or ruled grid dial, as well as on the fixed scanning grating, then the result is the digital autocorrelation function as shown in Figure 3.41d when the movable grating passes over the fixed grating. In this way, the sum of the corresponding bright

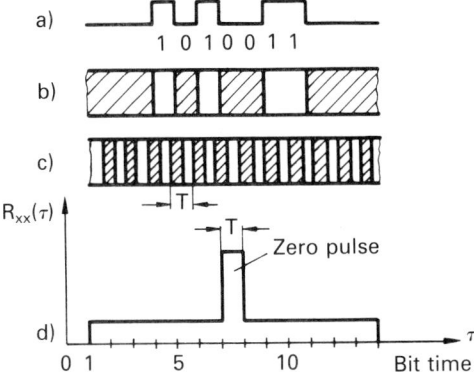

Figure 3.41   Pseudo-random 7 bit pulse train (a) and grid spacing (b) for generating a zero pulse (d) by means of autocorrelation function $R_{XX}$, and base grid (c).

fields is obtained [26]. The fact that this function, governed by the pseudo random distance between the grating lines, shows a characteristic maximum for a single position, means that a pulse of the desired position and width of a single increment can be generated by a trigger level. An essential criterion to ensure the quality of the random sequences used for the zero signal is, on the one hand, the absolute height of the maximum of the autocorrelation function, but on the other hand, the ratio between it and the ever present subordinate maxima. In Figure 3.41, the ratio of the main maximum to the subordinate maximum is 4:1, in Figure 3.42 7:2 and in Figure 3.43 11:5. In spite of the deteriorating ratio of wanted to unwanted signals, the Barker Code as shown in Figure 3.43 with bit 22 or longer pseudo-random pulse sequence must be selected, instead of random sequences with 7 of 15 bits, as the resolution increases. The autocorrelation function for the pseudo-random sequence 11100010010 and its complement are shown in Figure 3.43. Barker Codes and their autocorrelation functions play an important part in digital data transfer [27].

## 3.3   Electromechanical position sensing systems

One of the oldest processes still in use today, because of its simplicity, is the mechanical scanning of a scale with conductive and

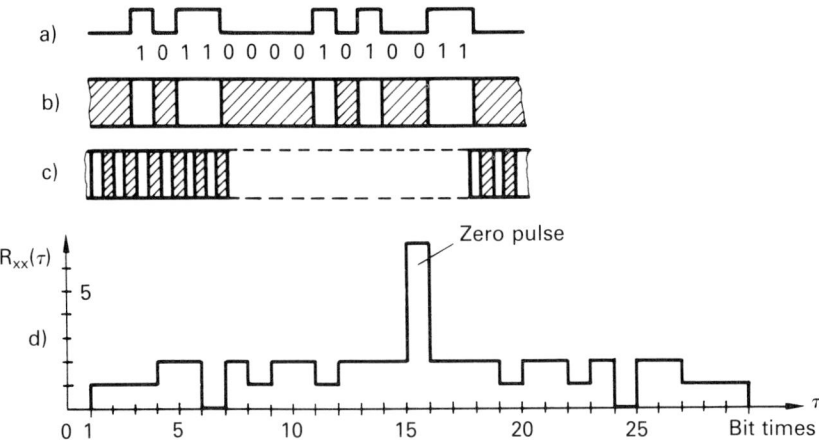

Figure 3.42   Autocorrelation function $R_{XX}()$ of a 15 bit random sequence.

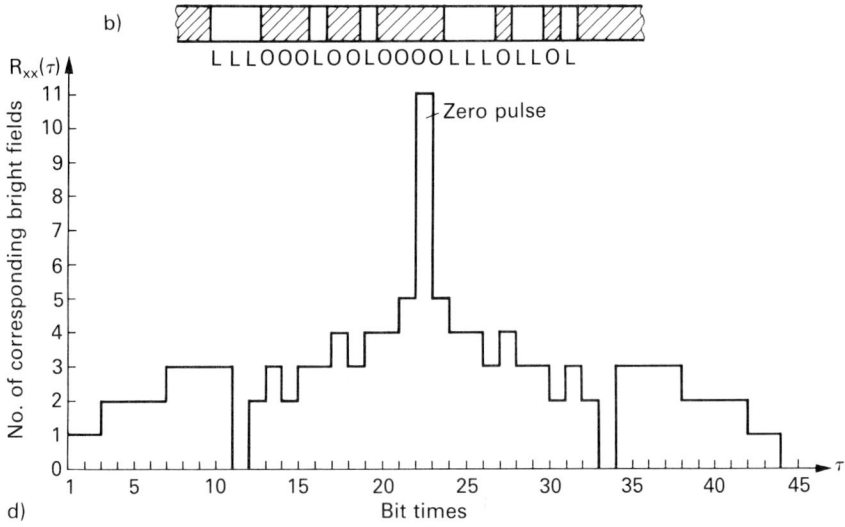

Figure 3.43   Using the Barker code to generate a zero pulse.

non-conductive zones. This makes possible, for example, the formation of a grating of gold conductive segments on an aluminium carrier over an isolating interface. All the segments are led to a common electrode which consists of a second continuous conductive strip. Both the grating strip and the continuous electrode are scanned using a brush. The measuring system is therefore an electromechanical switch which opens or closes an electrical circuit depending on position, Figure 3.44. A typical value for the voltage at which these contacts are required is 20 V. The maximum switching current is 2 mA [28]. As well as contact springs, contact pins are used which are pressed in by a separate spring. The advantage of this is that both the optimum contact and spring material can be selected for a particular task. All electromechanical pulse generators have the common disadvantage that the scales become worn. The sensing speed is considerably lower than is the case with no-contact scanning owing to the risk of contact bounce.

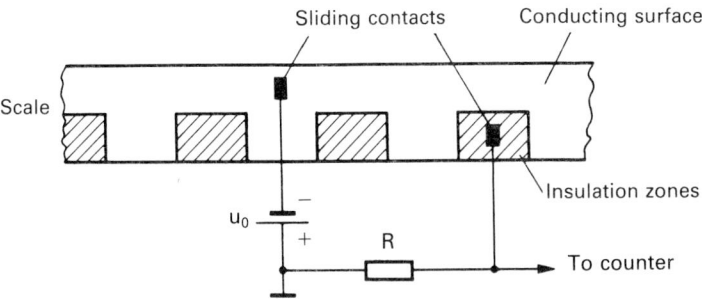

Figure 3.44   Electromechanical grid scale. $u_0$, voltage; $R$, resistance.

The limit of resolving capacity is reached at a segment width of approximately 0.1 mm. The advantages are compact construction and low cost.

## 3.4  Capacitive scanning

Mechanical wear of the scale is avoided when the capacitive scanning of a grating is carried out using a suitable backplate electrode. The main disadvantage of this process is the high internal resistance of the signal source and the low signal amplitudes which result. The limit for the resolution of capacitive scales is the same order of magnitude as that of electromechanical systems. In spite of numerous patents in this area capacitive displacement sensors have not achieved any significance.

## 3.5   Carrier frequency scanning

In the case of an angle encoder which uses carrier frequency scanning, the measurement base is a toothed disc or wheel and the transducer records the number of passing teeth. A characteristic of the scanner is that it contains a frequency-determining element of an oscillating circuit, e.g. a coil. If this coil is near a magnetic material, its inductivity is altered and with it the frequency of the oscillating circuit. The two different frequencies can be defined as logic 0 and logic 1 [29]. The principle corresponds closely to that of proximity detectors. It is characterized first and foremost by its great mechanical robustness, but its resolution is low.

## 3.6   Electromagnetic pulse generator

There are two basic processes used with electromagnetic incremental transducers. In the first, scales are made of magnetic material of high coercive force in which areas of high magnetic flux density alternate with those of low flux density. Ferrite cores with an almost rectangular hysteresis loop are useful for sensing. The ferrite cores have a read coil and a sense coil. The sense coil is energized using a frequency between 20 and 200 kHz. As long as the ferrite core is positioned over a non-magnetic area, the hysteresis loop is governed by fully advanced control from the sensing current. A maximum voltage, which can be defined for example as logic 1, occurs in the sense coil. As soon as the ferrite core arrives over an area of high flux density it is brought to saturation point by additional magnetization and the change in flux emanating from the sensing current and with it the voltage in the read coil becomes very small, Figure 3.45. This state is defined as logic 0 [28;30]. In the second scanning process Hall probes are used for sensing [31]. The Hall effect is particularly apparent in the indium antimonide (InSb) and indium arsenide (InAs) III–V compounds. If a thin rectangular lamina of these compounds is placed in a magnetic field and a constant current is allowed to flow through this lamina, an electrical potential is then generated across the current conduction direction which is proportional to the magnetic induction and the auxiliary

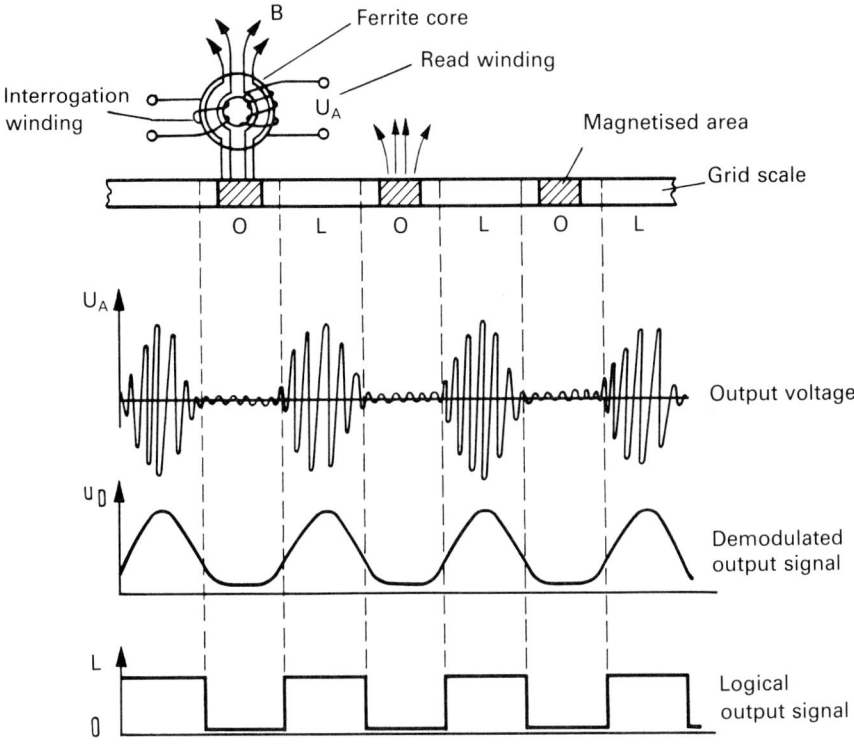

Figure 3.45   Scanning a magnetized grid scale.

current *I*, Figure 3.46. The potential is referred to as the Hall potential $U_H$. The relation for this is:

$$U_H = K_H IB \qquad (40)$$

where *B* is the magnetic induction, *I* is the auxiliary current and $K_H$ is the Hall constant. The Hall constant $K_H$ is a property of the material. The auxiliary current *I* must be kept constant for all measurements. The scale grid is actually characterized by alternating magnetization, i.e. north and south poles alternate. Depending on the direction of the magnetic field, either a positive or a negative Hall potential is generated. Integrated comparators then convert these potentials into binary signals. As with the electromechanical scanning processes, the limit for resolution with the electromagnetic processes is approximately 0.1 mm as signal amplitudes diminish as gratings become finer. Optical gratings are used for very fine divisions.

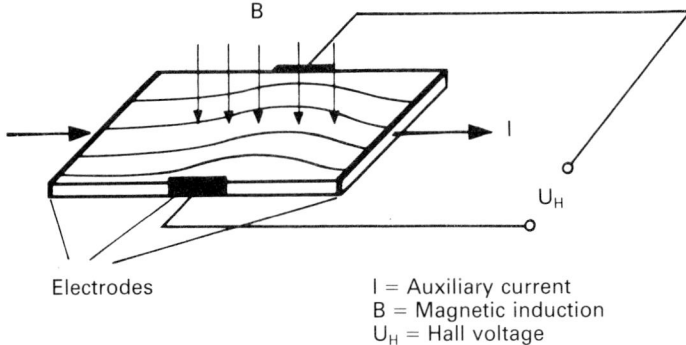

Electrodes

I = Auxiliary current
B = Magnetic induction
$U_H$ = Hall voltage

Figure 3.46   Hall generator.

## 3.7   Photoelectric position sensing systems

Very sharp optical images can be projected over the short distances required using specified light/dark transition points, as the wavelength of the light is small even in comparison with the divisions of very fine grids. In contrast, electromagnetic and capacitive scanning methods do not have such distinct transition points. A distinction is made in scanning optical scales between the transmitted light and incident light methods. In addition, a distinction should also be made between amplitude and phase modulation of the light. In the transmitted light method glass scales are used which have a grid consisting of alternating transparent and opaque fields. Figure 3.47 shows such a grid scale with a 1 : 1 division ratio. The grid is illuminated with parallel light emanating from a point source $L$ with condenser lens $K$. A slit diaphragm or a second grid, behind which the photodetector, e.g. a photodiode, is located, is placed directly in front of the grid scale in order to begin scanning operations. The distance between the scale and the scanning plate (slit or grid) is a few tenths of a millimetre. The luminous flux to the photodetector and therefore its electrical signal is a function of the ratio of the width $\delta$ of the scanning slit to the division of the grid, Figure 3.48. It is desirable to have a path in which the transition points from the 0 condition to the 1 condition and vice-versa occur in an infinitely brief period, Figure 3.48. The width of the scanning slit and the grid division are equal, Figure 3.48b, which happens when the divisions are very fine. Ideally a triangular path is produced for the luminous flux, but in practice its shape usually appears more sinusoidal. If under the same illumi-

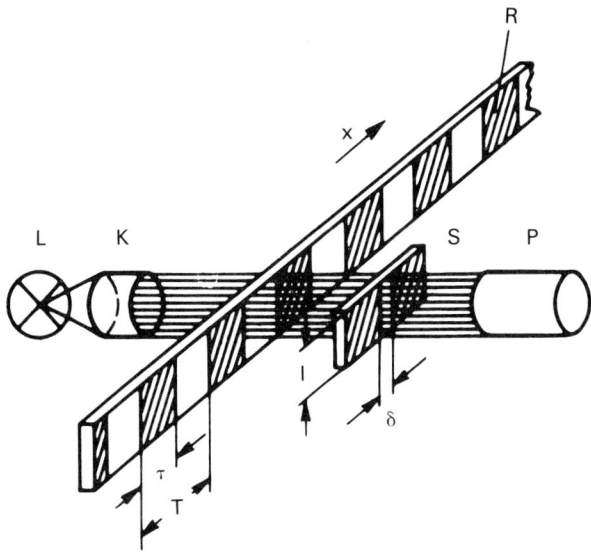

Figure 3.47   Photoelectric scanning of a grid scale. L, Light source; K, condenser; $T$, grating period; $\delta$, width of scanning slit; $\tau$, grid division; $x$, direction; S, slit diaphragm; $R$, grid scale; P, photodetector; $l$, length of slit.

nation the receiving slit is reduced in size the amplitude of the luminous flux decreases, Figure 3.48c. The curved shape, however, again approaches the ideal path. Where the slit width is as in Figure 3.48c the original amplitude can be reproduced if the luminous intensity of the light source is increased in inverse proportion to the slit width, Figure 3.48d. The most favourable conditions for the photodetector signal are therefore created when the slit width is as small as possible and the luminous intensity is as high as possible. High luminous intensity, however, means equally high heat losses, which are not only harmful to the photodetector and its amplifier but with the heating of the scale and the guide may also result in unacceptably large errors due to thermal expansion. In high-resolution systems use is therefore made of the possibility of separating the light source and the scanning device by illuminating the scale by means of a fibre-optic cable in series with a heat filter. Such assemblies are referred to as cold light sources. As the grid lines in high-resolution systems appear small against the dimensions of the photodetectors the scale is scanned simultaneously at several adjacent points within the space of a grid division. The luminous flux increases in proportion to the number of scanning slits. In other words, the scale grid is scanned

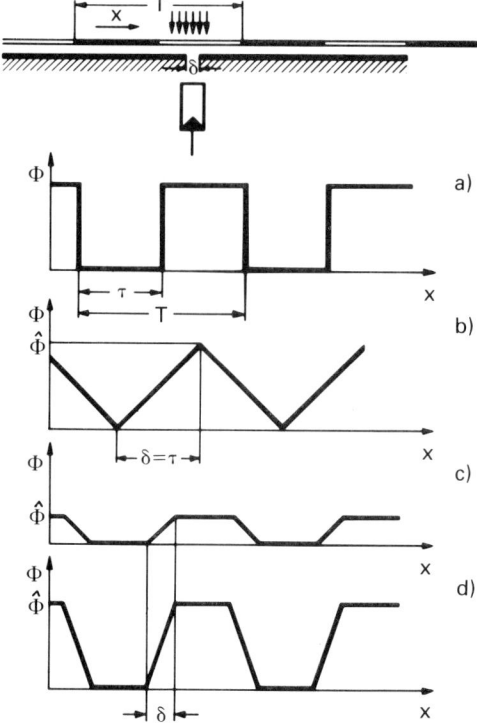

Figure 3.48    Luminous flux during the displacement of an optically scanned grid scale. $\varphi$, luminous flux; $\tau$, grid division; $T$, grating period; $\delta$, width of scanning slit; $x$, direction.

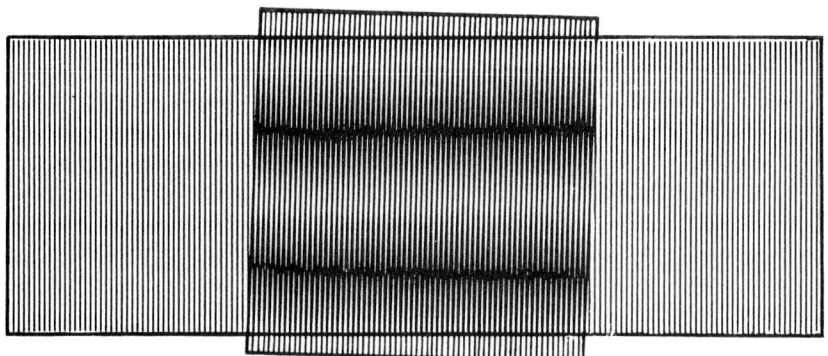

Figure 3.49    Producing moire fringes by crossing two gratings.

using a second grid with the same size of divisions. If the two grids are moved towards each other, the luminous flux changes periodically with the grid division as the gaps and lines of both grids are covered alternately. In addition to the effect of increased sensitivity at specified illuminations, the use of a scanning grid produces greater accuracy. It can be assumed that each scale is affected by statistically distributed errors during the dividing procedure. When several grid lines are being scanned simultaneously the results are therefore averaged. If the grid scale and the scanning grating are at a slight angle to each other, moiré fringes are produced in the transmitted light, Figure 3.49. The spacing between these fringes is a function of the angle between the two gratings. If the scanning grating is moved horizontally towards the scale, the moiré fringes will move upwards or downwards depending on whether the scanning grating is moving to the right or to the left. If such a moiré fringe is scanned using a photodetector located behind a horizontal slit, and the scanning grid is moved horizontally, then the luminous intensity is altered on the photodetector until it becomes near-sinusoidal. The advantage of the moiré effect is that the space between two adjacent light or dark sectors is magnified in relation to the original grid. In order to generate two 90° phase-shifted signals, two photodetectors must be placed one on top of the other. The space between them must be equal to a $\frac{1}{4}$ period of the moiré fringes. By using the moiré effect to magnify the original grid, more favourable conditions are produced for the space between the photodetectors [33;34].

### 3.7.1 Radiation sources

Incandescent lamps, light-emitting diodes, laser diodes and lasers are used as light sources in modern angle and displacement measuring systems. At present the low voltage neon-glow lamp is only used in indicating elements in the form of plasma indication.

### 3.7.1.1 Incandescent lamp

In spite of the penetration of light-emitting diodes the incandescent lamp is holding its own in many applications. This is partly due to the development of lamps with a life of up to 100000 h, with the result that a powerful argument against the use of the incandescent lamp, its limited life, is now only true in certain cases. The following equation is used for the operating life $L_B$ of a tungsten lamp in relation to the operating voltage $U_B$ used:

$$\frac{L_B}{L_N} = \left(\frac{U_N}{U_B}\right)^{13,14} \tag{41}$$

In this relation $L_N$ is the nominal lifespan and $U_N$ the nominal operating voltage. The exponent 13 applies to vacuum lamps and 14 to gas-filled lamps. This relationship is accurate to within ±20% of the nominal voltage. Where deviations from the nominal voltage are greater, the values calculated can only be regarded as orders of magnitude. Figure 3.50 shows the characteristic lines of incandescent lamps in the interesting range between −20% and +20% of the nominal voltage. The expression for the light current shows that it roughly increases by the fourth power of the voltage:

$$\Phi \sim KU^4 \tag{42}$$

The lamp current itself increases less than proportionally with voltage, as the resistance in the tungsten coil increases with increasing voltage and therefore increasing temperature. Increasing the voltage by 1% results in a rise in current of between 0.5% and 0.6%. In order

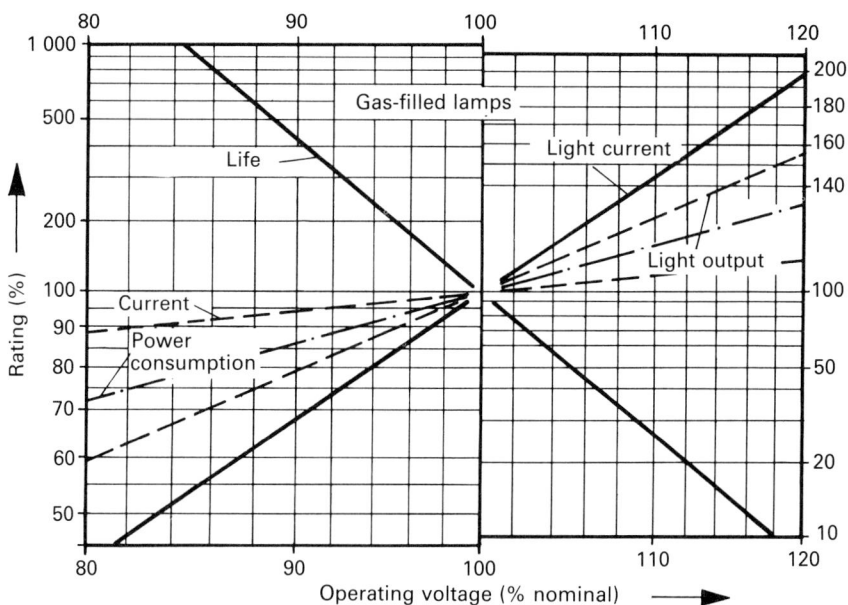

Figure 3.50   Alteration of the operating data for incandescent lamps with voltage variations around the rated voltage.

to clarify the effect of the operating voltage, let us choose two numerical values: lowering the operating voltage by 5% means a doubling of lamp life and lowering it by 15% means a tenfold increase! It is therefore possible, by means of a relatively small reduction in operating voltage, to achieve a much longer lamp life in relation to the nominal data with only slightly reduced light current. In practice the majority of incandescent lamps used for measuring purposes are operated on low voltage. Care is required, however, with halogen lamps, as the iodine filling means that other conditions apply.

The result of gas filling is that the rate of evaporation of the tungsten in the filament is lowered because of gas pressure, which means that higher operating temperatures are possible. Although losses due to heat dissipation caused by the gas increase roughly in proportion to the temperature $T$, the specific consumption nevertheless increases with increasing temperature as the radiation power rises by between $T^4$ and $T^5$. The actual purpose of the iodine fillings is, however, to prevent the bulb being blackened by evaporated tungsten. In the lamp bulb the iodine combines with the tungsten to form tungsten iodide gas, provided that the temperature of the glass wall is more than 250°C and less than 1450°C. At temperatures above 1450°C, which occur at the filament and in the surrounding layer of gas, the tungsten iodide separates once more into tungsten and iodine. The drop in temperature between the incandescent body and the lamp bulb gives rise to a continuous cyclic process in which the tungsten from the bulb is transported back to the coiled filament. Unfortunately it is not always deposited in the same place, with the result that halogen lamps have only a finite lifespan. If halogen lamps are heated from below the cyclic process can be interrupted, but this can be compensated for to some extent by reduced evaporation rates. Of greater significance is the fact that a specific amount of iodine is added according to the operating temperature for which the lamp has been designed. If the operating temperature is lowered, this relation is no longer correct, so that the $U^{14}$ law must be used with the greatest care outside the ±20% range. Another danger is that a recrystallization may take place in the tungsten where temperatures are too low, resulting in brittleness and consequently greater susceptibility to shocks.

Figure 3.51 shows on a linear scale the change in lamp resistance, current intensity and luminous flux caused by the operating voltage. It can be clearly seen that the cold resistance of the lamp is considerably less than the resistance during operation at duty rating.

This means that the starting current in incandescent lamps can be up to ten times greater than the operating current. The result of this is that a lamp will usually burn out when it is switched on. This overload can be avoided by limiting the current, thereby also increasing the lifespan. Achieving a specific lifespan is only one of the criteria for choosing a lamp. Other aspects to be considered are shape, luminous intensity and colour temperature. With regard to shape it must be said that as far as possible small lamps are of course preferred. This requirement has been met by the development of low-voltage lamps with ratings of 6 or 12 V, halogen-filled and with quartz bulbs. The short coiled filament delivers a point light source with a high degree of accuracy, besides which the filament is considerably more stable mechanically than is the case with lamps having a higher working voltage. The coiled filament has greater thermal inertia and in many cases may therefore be powered using alternating current without any fluctuation in the luminous intensity becoming noticeable to the point of interference. Adapting the luminous intensity to suit the appropriate detector must always be seen in the context of spectral adaptation. This means that colour temperatures is an important quantity. Although the colour temperature describes only a physiological impression of colour, it does given an exact description of the radiated spectrum when used in conjunction with a thermal radiator such as tungsten. Figure 3.52 shows the spectral distribution curve of the radiation flux with a black, grey or tungsten body, where tungsten can be regarded as a grey body. The spectral distribution curve is explained in Wien's displacement law:

$$g_{max} \times T = 2.898 \times 10^{-3} \, [\text{m K}] = \text{constant} \tag{43}$$

This says that the maximum amount of radiation is displaced into the lower wavelength region of the spectrum as the temperature rises.

### 3.7.1.2  Light-emitting diode

Unlike incandescent lamps, where light emissions relies on a thermal effect, light-emitting diodes produce light by the internal photoelectric effect in semiconductor materials. Internal photoelectric effect means the recombination of charge carriers in the p–n junction, which occurs during radiant emission. The wavelength of the radiation emitted by the light-emitting diodes depends on the semiconductor material used and its doping. Common materials used are gallium arsenide (GaAs), gallium arsenide phosphide (GaAsP)

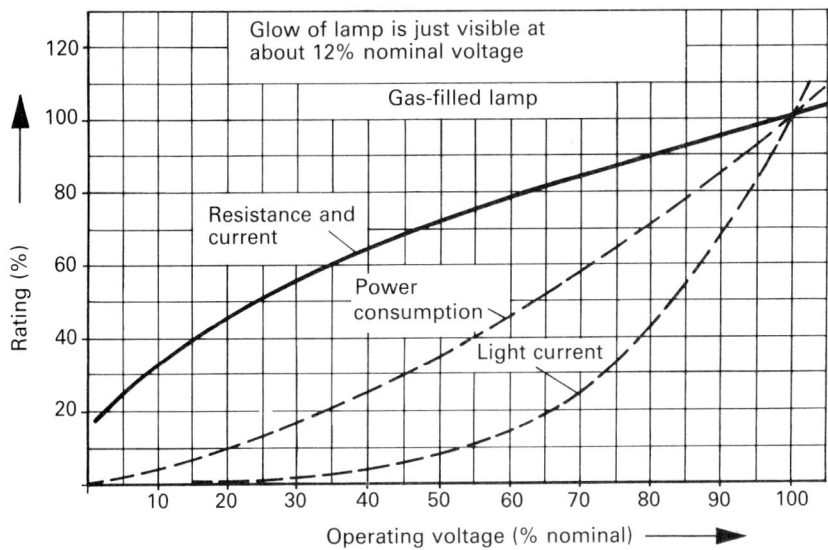

Figure 3.51 Characteristic curves of incandescent lamps at less than rated voltage.

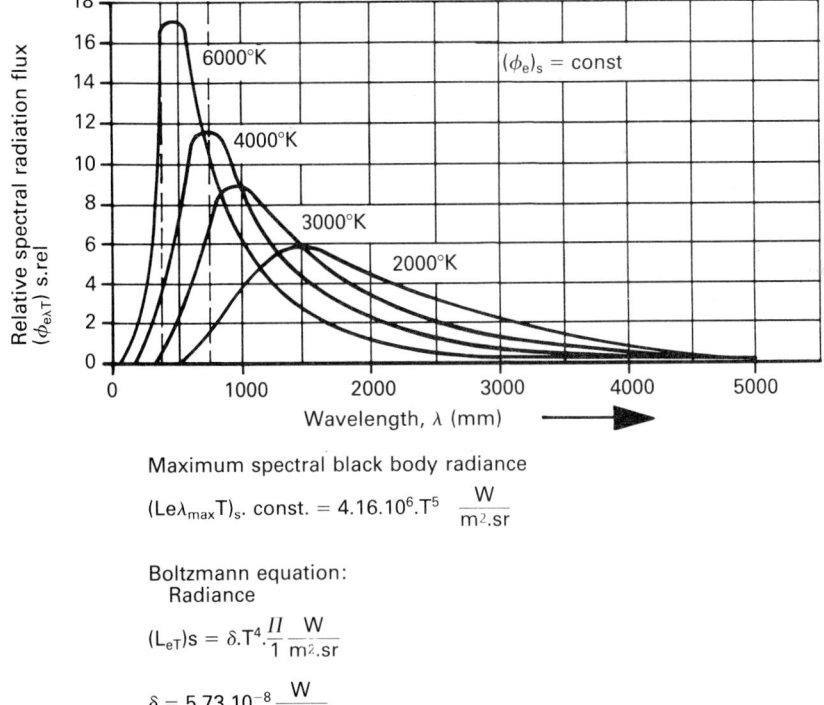

Maximum spectral black body radiance

$$(Le\lambda_{max}T)_s. \text{ const.} = 4.16.10^6.T^5 \quad \frac{W}{m^2.sr}$$

Boltzmann equation:
Radiance

$$(L_{eT})s = \delta.T^4.\frac{II}{1}\frac{W}{m^2.sr}$$

$$\delta = 5.73.10^{-8}\frac{W}{m^2.k_4}$$

Figure 3.52 Significant relations for temperature radiators.

and gallium phosphide (GaP). GaAs diodes emit light in the infrared region. The spectrum ranges from 0.88 to 0.94 μm, attaining a maximum of 0.9 μm, Figure 3.53. The emission spectrum matches very well the spectral sensitivity of silicon photodetectors, Figure 3.54.

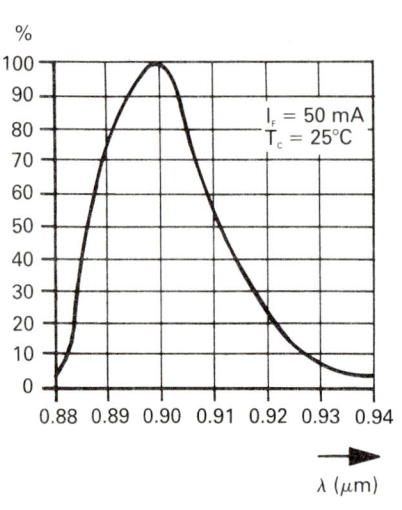

(a)    Relative/radiant intensity

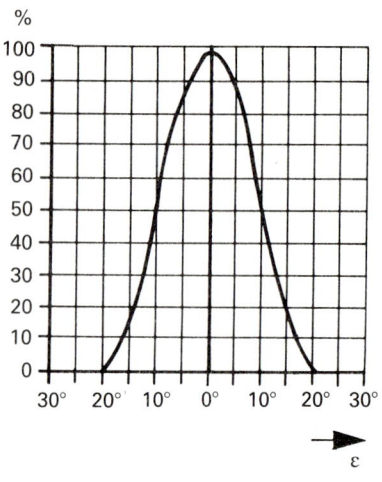

(b) ε   angle between diode optical axis and direction of measurement

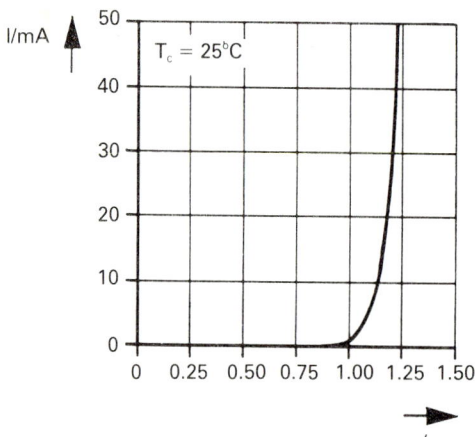

(c) Current–voltage characteristic in direction of transmission

Figure 3.53   Characteristic curves of a GaAs diode.

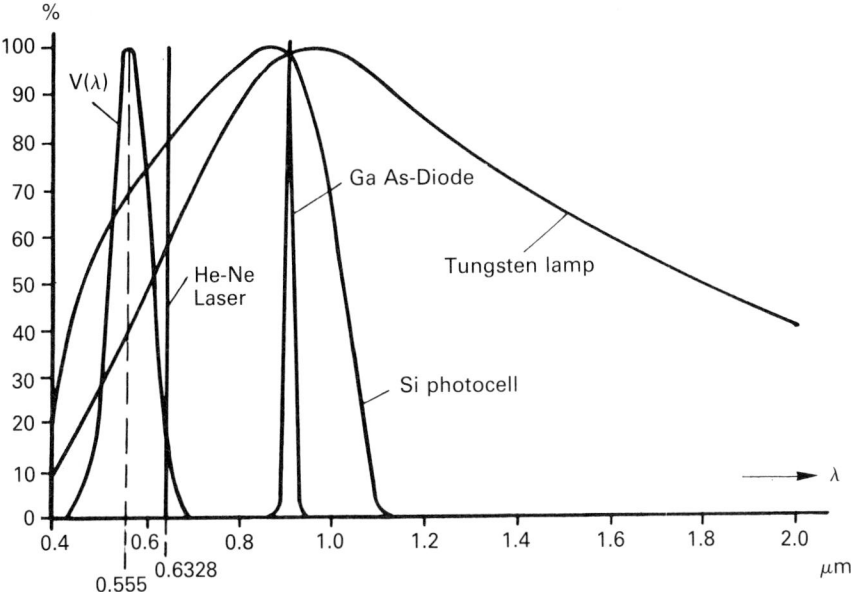

Figure 3.54 Spectra of different radiation sources and sensitivity of Si photodetectors and the human eye.

GaAsP diodes emit red or yellow light and red, yellow or green emitting diodes can be made using GaP. As radiation is produced by recombination of charge carriers, light-emitting diodes, often abbreviated to LED, produce what is known as 'cold' light. The active range of light-emitting diodes depends on ambient temperature, which is why no measures need be taken to dissipate heat. With a conducting-state voltage between 1 and 2 V and a current between 5 and 10 mA in continuous operation, power loss ranges between 30 and 50 mW, Figure 3.53. Reduced power is not the only advantage of LED over incandescent lamps. A key factor is that the radiation follows the current almost free of lag. Build-up and fall times can be counted in nanoseconds, which means that modulation of luminous flux is possible well into the gigaHertz region.

The cut-off frequencies which can be achieved depend to a large extent on the type of doping. For example, zinc-doped GaAs diodes are faster than silicon-doped diodes. Typically, the life of light-emitting diodes is 100 000 h, and the end of their life is not determined by total loss of power but by reduction of their radiation capacity to half of the initial value. The luminous intensity mostly changes linearly with the intensity of current, so that analogue data transfer is also possible with light-emitting diodes. It should, however, be noted that the threshold voltage is highly temperature-dependent, which means that it is best if the power supply is constant current. But even using constant current, radiation capacity will change with temperature. It will be reduced by about 1%/°C, approximately linearly with the temperature.

In addition, the following advantages of light-emitting diodes should be mentioned: small size and insensitivity to shock and vibration stresses [36;37].

### 3.7.1.3   Laser

The light from a laser has advantages over all other light sources in that it is strictly monochromatic, has a steady amplitude and is coherent in time and space. Monochromatic means that oscillations of only a single frequency are produced. In order to gain an understanding of these characteristics, the physical interrelationships in the gas laser should be considered. The starting point is the Bohr atomic model, which describes the energy states of electrons in a gas atom. According to this theory, electrons move in circular paths about the nucleus, and each path corresponds to a specific energy state of the electron. The further an electron is from the nucleus, the greater its energy content. By absorbing energy, an electron can make the transition to an orbital path more distant from the nucleus, but the energy can only be absorbed in discrete values which correspond to a multiple of Planck's quantum of action $h$. Excitation is made possible by the absorption of electromagnetic radiation, by heat or by shocks in a gas discharge. If an electron is raised from its lowest energy level, where it is situated without absorbed external energy, to a higher level, then it will remain in this energized state for between $10^{-8}$ and $10^{-3}$ s. It will then fall back to its normal state, emitting energy in the form of radiation. The frequency of the radiation will depend on the difference in energy between the two levels of the electron concerned. As an energized atom falls back spontaneously into the normal state no specific phase relation exists between the individual wave trains of a radiating body, e.g. an incandescent lamp. Besides, each wave train has only a finite length. The laser effect is thus made possible by the fact that an energized atom can be caused to fall back to the normal state not only spontaneously but also by external influences, e.g. by interaction with an external electromagnetic field (stimulated emission). The frequency $f_L$ of the stimulant radiation should therefore correspond to the energy difference between the two levels $E_a$ and $E_b$ of the energized electron:

$$f_L h = E_a - E_b$$

where $h$ is Planck's quantum of action.

As a further requirement for the laser effect, a suitable amount of energy must be absorbed to ensure that the higher energy level is inverted so that the portion of irradiated energy contributing to the stimulant emission is greater than the absorbed portion. Only then is light amplification possible. In the He–Ne laser, inversion of the upper energy levels is brought about by a discharge in gas. The active material is neon, which has two energy levels, and the energy difference between them gives rise to radiation where the wavelength $\lambda$ = 0.6328 $\mu$m ($f_L$ = 4.7 × $10^{14}$ Hz). Although helium has what is known as a metastable energy level in the same place as neon, it is not possible to vacate this level by means of stimulant radiation. Inelastic collisions between energized helium atoms and neon atoms may, however, bring about a transfer of energy, thereby producing further energized atoms. The helium is therefore used only to provide additional excitation, creating an inversion state in the neon. The He : Ne mixture ratio is about 10 : 1. A high energy density of the stimulant radiation is required to achieve a multiple of the power of the stimulant emission over the ever present spontaneous emission (noise). This energy density is achieved by incorporating a gas discharge vessel in an optical resonator consisting of two mutually opposed concentrating reflectors, Figure 3.55 [38]. One reflector is almost 100% reflective whereas the other has a transparency between 1% and 2%. When the laser is switched on, a non-directional and random radiant emission is first produced by collision ionization in accordance with the spontaneous emission of neon atoms. Most of this radiant emission is lost, except for the small portion which falls on the concentrating reflector parallel to its axis and is then reflected back into the gas discharge chamber. This stimulates further radiant emission in the same phase and with equal wavelength by causing energized neon atoms to emit energy. As the stimulated radiation propagates parallel to the path of the energizing radiation, and therefore parallel to the axis of the resonator, this process increases expotentially until equilibrium between excitation energy and energy losses has been achieved. The gas discharge acts like an amplifier for light of a specific wavelength, whereby reaction occurs through the optical resonator. The length of the resonator determines the exact frequency at which the laser vibrates, as the spectrum line has a width of

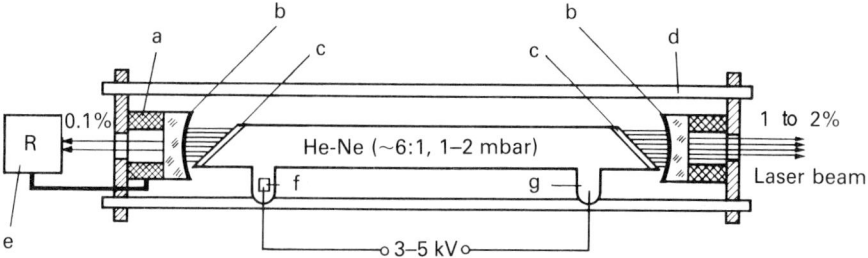

Figure 3.55   Construction of an He–Ne laser. a, Electrostrictive element; b, spherical mirror; c, Brewster window; d, quartz or invar connecting rods; e, frequency stabilization; f, cathode; g, anode; R, automatic control.

about 1.6 GHz. This spectrum line, which at first glance seems extraordinarily wide, means a relative frequency uncertainty of 2.7 × $10^{-6}$. This is principally determined by the interaction between atoms and by the Doppler effect. As a result of the continuous motion of gas molecules, various frequency displacements occur according to the speed and direction of motion of the various atoms. External influences, such as a temperature, would, however, change the wavelength unacceptably, even with a resonator. For linear measurement, however, a relative frequency constancy between $10^{-7}$ and $10^{-9}$ is needed, which can only be achieved by additional stabilization measures. These achieve stabilization by mechanically changing the space between the two resonator reflectors. To this end, one of the two reflectors is mounted on an electrostrictive element (e.g. made of piezoelectric ceramics) the length of which can be altered by means of an electric field. Two optical effects are predominantly used as a criterion for adjustment: one is the fact that the output of the laser beam is at a minimum when the natural mode is accurately tuned to the middle of the spectrum line (lamp-dip) [39]; the other is the Zeeman effect. This can usefully be described in greater detail [38;40]. Under the influence of a longitudinal magnetic field, the light from the laser is split into two opposing circularly polarized components. These two components are of equal intensity if the resonator frequency $f_o$ is in the middle of the original neon line, Figure 3.56.

Adjustment by means of a Zeeman absorption cell is described in [41]. A small fraction of the beam is decoupled on the back of the laser, circularly polarized and then emitted by a Zeeman absorption cell. This Zeeman absorption cell consists of a neon cell, to which a longitudinal magnetic field is applied which splits the absorption curve into two symmetrical curves, one of which controls only the

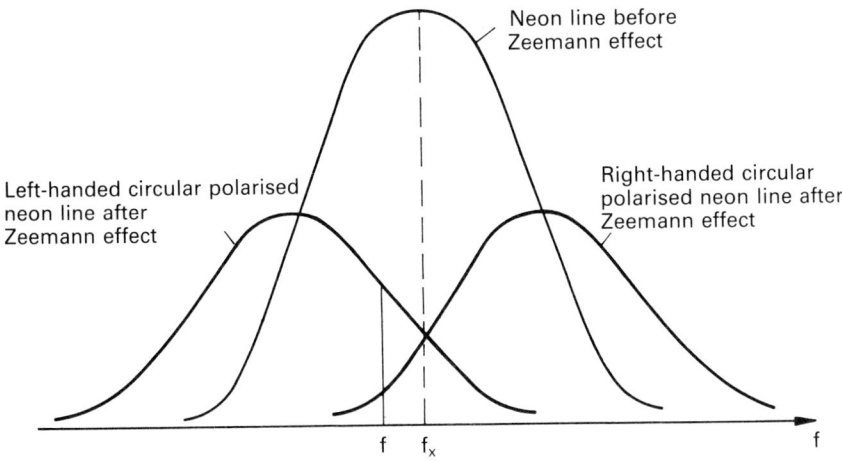

Figure 3.56   Zeemann effect on neon line.

light circularly polarized to the right and the other controls only the light circularly polarized to the left. This means that the absorption curves correspond to the spectral divisions, Figure 3.56. If the orientation of the magnetic fields is reversed the correspondence returns to the polarizing directions. If the magnetic field is energized using an alternating voltage the absorption varies periodically as the frequency of the laser light deviates from frequency $f_0$. In this case, a photodetector placed behind the absorption cell supplies an alternating voltage. A downstream phase-sensitive rectifier can be used to detect the direction of the deviation. Finally, the output signal behind the rectifier can be used to control the piezo crystal and thus to adjust the length of the resonator. In the short term, the relative frequency fluctuations $\Delta f/f_0$ can be reduced to $2 \times 10^9$. Long-term changes are brought about by fluctuations of $f_0$, which can, however, be kept below 100 kHz per day of operation.

## 3.7.2   *Photodetectors*

There are two groups of photodetectors, one of which uses the internal and the other the external photoelectric effect. In the case of the internal photoelectric effect the action of light is used in semiconductors to produce charge carriers which increase conducting capacity. In the case of the external photoelectric effect electrons are liberated from the surface of solid bodies as a result of radiation exposure.

Examples of the internal photoelectric effect include the photovoltaic cell, photodiode, photoresistor, phototransistor and photothyristor and those of the external photoelectric effect include the gas-filled photoconductive cell, vacuum photoconductive cell, secondary-emission multiplier and some TV cathode ray tubes. To the above-mentioned examples can be added a number of special forms such as the p–i–n photodiode and the avalanche photodiode, which are designed for specific applications. Figure 3.57 shows some typical voltage–current characteristics, and Figure 3.58 examples of sensitivity and response time constants, of photodetectors. The criteria which are of vital importance in angle and displacement measurement are: structural dimensions, range of spectral sensitivity, simplicity of the amplifying circuit and, last but not least, cost. In practice the selection is therefore reduced to the photovoltaic cell, the photodiode and the phototransistor. These are semiconductor components with one or two depletion layers in which charge carriers are produced by radiation, thereby effecting an increase in the off-state current. Spectral sensitivity, which can be at its maximum in the visible nor non-visible region, does not depend on the structure of the particular semiconductor component but solely on the semiconductor material used. The photodetectors used in angle and displacement measurement processes are almost exclusively made of silicon, germanium photodetectors being used only in special cases. The reason for this is that the relative spectral sensitivity distribution of silicon is not only well suited to the emission spectra of the radiation sources which are most important for this application (tungsten filament lamps, GaAs diodes and He–Ne lasers), but it also has sufficiently high values in the region of visible radiation (see Figure 3.54). In addition the high infrared sensitivity of germanium causes interference in many applications, giving rise to a considerable dark current due to the influence of temperature.

The basic structures of the photodiode and the photovoltaic cell are identical as they both have a p–n junction. Any photodiode can be operated as a photovoltaic cell and vice versa. Figure 3.59 shows this using as an example the voltage–current characteristics of a photosensitive p–n junction [42]. Quadrant I shows the diode operating in the conduction direction. The characteristic curve of the unlit diode is displaced in the direction of the negative axis of current as the illumination $E$ increases. Quadrant I, however, is of no significance for applications requiring a light-activated component.

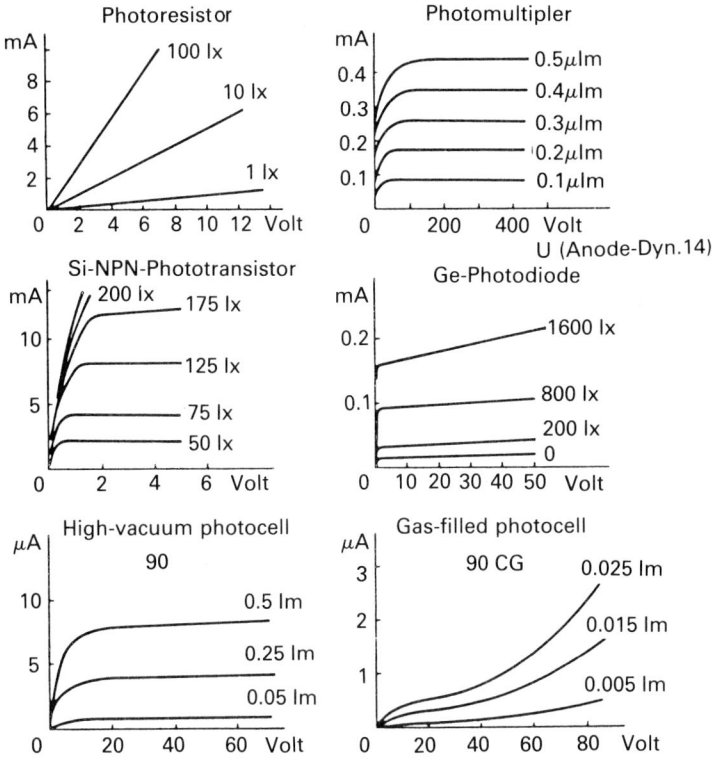

Figure 3.57   Current–voltage characteristics of different photodetectors.

It is Quadrant III (operating as a photodiode) and Quadrant IV (operating as a photovoltaic cell) which are of interest. If operated as a photovoltaic cell, pairs of charge carriers are produced in the space-charge region created by the p–n junction by the action of light resulting from the internal photoelectric effect. The electric space-charge field draws the charge carriers out of the p–n junction and produces a short-circuit current in the external circuit. This short-circuit current is strictly proportional to the illumination $E$ up to illuminations of over $10^5$ lux. If the external circuit is provided with a load resistance $R_a$ a photoelectric voltage is built up across the p–n junction.

   In open-circuit operation, the charge carriers generated by light absorption cannot be carried off outside. This alters the space-charge conditions and an open-circuit voltage is created at the external terminals. This rises logarithmically with illumination. As the space-charge region degrades at a specific illumination, the open-circuit voltage reaches a saturation value. Unlike the short-circuit current which rises linearly with illumination to reach very high illuminations, the open-circuit voltage is restricted to a maximum value, Figure 3.60. When

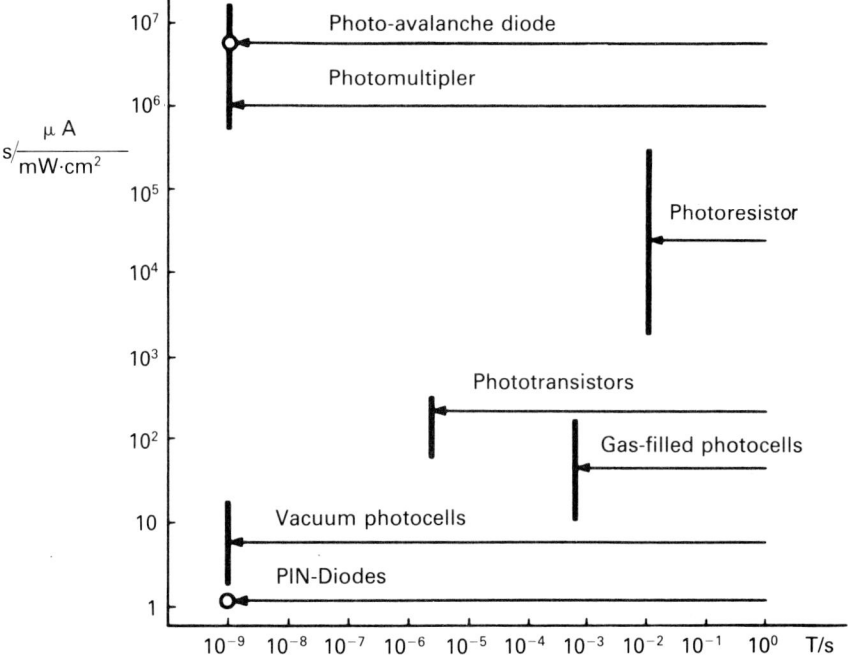

Figure 3.58   Sensitivity and switching time constants of photodetectors.

operating as a photodiode (Quadrant III) a voltage in the barrier direction is applied to the p–n junction. In order to obtain a sufficiently large signal at low illuminations and therefore low photoelectric currents, a maximum load resistance $R_a$ and a maximum voltage are required. As the capacitance of the depletion layer decreases with rising barrier voltage, the cut-off frequency rises with the barrier voltage at a specified load resistance [43]. As the cut-off frequency, however, is at the same time in inverse proportion to the load resistance, it is sometimes necessary to seek a compromise between the desired signal amplitude and the required cut-off frequency.

Although any photodiode can be operated as a photovoltaic cell and vice versa, it is useful to make a distinction as, depending on the desired application, certain characteristics can be 'grown'. For example, although a large area depletion layer means a high degree of photosensitivity, it does at the same time give rise to a large depletion layer capacitance and a high dark current.

In photodiodes the depletion layer is kept small to obtain a low dark current and a high barrier voltage. This limits the sensitivity of photodiodes. In the case of photovoltaic cells, which are operated in the conducting direction, the photosensitivity should, in contrast, be

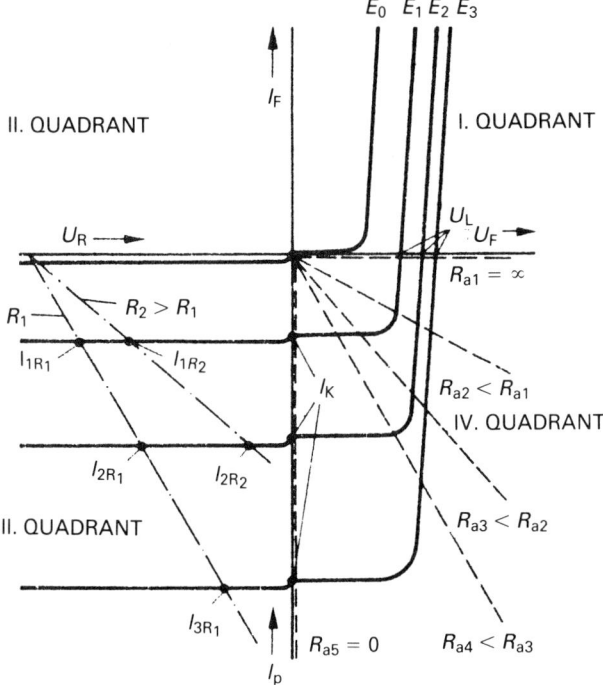

Figure 3.59   Current–voltage characteristics of a semiconductor photodiode with resistance lines. Third quadrant: operating range of the photodiode. Fourth quadrant: operating range of the photovoltaic cell. Parameters: $E$, illumination ($E_0 = 0$, $E_3 . E_2 . E_1$). $U_R$, $U_F$, cut-off voltage applied or forward voltage occurring; $I_R$, $I_F$, cut-off or forward current; $U_L$, no-load voltage; $I_K$, short-circuit current; $R_a$, external resistance ($R_{a1} = 0$, no-load; $R_{a5} = 0$, short-circuit).

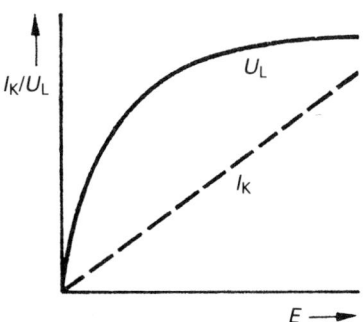

Figure 3.60   Short-circuit current $I_K$ and no-load voltage $U_L$ as a function of illumination $E$.

as great as possible, which results in a large area depletion layer. When in photovoltaic mode there is no dark current. The barrier voltage of photovoltaic cells is thus considerably lower (between 1 and 20 V) than is the case with photodiodes (between 10 and 100 V), but this is of no importance in this operating mode. The temperature dependency of light-activated voltages and currents plays an important part in position detecting systems.

Thus the open-circuit voltage of photovoltaic cells diminishes by approximately $2.3\,\text{mV}\,\text{K}^{-1}$. At about $400\,\text{mV}$ saturation voltage this is already over $0.5\%\,\text{K}^{-1}$. In contrast, the short-circuit current has a temperature coefficient of only $0.1\%\,\text{K}^{-1}$. Response times are another characteristic quantity. In photovoltaic cells in short-circuit mode they are limited by the propagation time of the charge carriers and in open-circuit mode by the depletion layer capacitance and the internal resistance. In open-circuit mode they are also highly dependent on illumination. Typical values for switching times can be given as a few microseconds for short-circuit mode and between $50\,\mu\text{s}$ and $10\,\text{ms}$ for open-circuit mode, depending on the illumination.

The phototransistor basically consists of two diodes, the emitter-base diode and the collector-base diode, one of which operates in the conducting direction and the other in the barrier direction. The collector diode is operated as a photodiode. From the functional point of view the phototransistor is a conventional transistor with a photodiode between its base and collector.

In principle, the phototransistor can therefore be reduced to the function of a photodiode with a donwnstream transistor in which the radiation-controlled off-state current of the diode is amplified by the transistor. Phototransistors are usually operated in common emitter mode in which the base terminal is frequently not led through. The sensitivity of phototransistors brought about by the internal transistor is between 100 and 500 times higher than that of photodiodes. At the same time, however, the dark current of the photodiode is being amplified. A further disadvantage of the phototransistor is the considerably increased switching times compared with photodiodes and photovoltaic cells. For reasons of sensitivity a relatively high depletion layer capacitance is available, the efficiency (time constant) of which is further increased by the current amplification factor ß. To summarize, it can be said that the use of photovoltaic cells in short-circuit operation is recommended where a high degree of thermal stability and a high cut-off frequency are required. As the signals are usually to be found in the A region an appropriate sum has to be spent on downstream amplifiers. With photodiodes, and in particular with phototransistors, greater sensitivity is obtained at the cost of a not inconsiderable amount of dark current. Moreover, in phototransistors the relatively high switching times can also cause interference.

Special forms of photodiode include the p–i–n photodiode, the avalanche photodiode and the Schottky photodiode. The following operating characteristics are to be noted. P–i–n photodiodes have switching times ranging from picoseconds to nanoseconds. They also have a very low dark current (order of magnitude 1 A). They are therefore suitable as receivers of light-induced pulses in the nanosecond region (laser) and for transmitting signals with high-frequency modulated light (up to about 50 MHz). In the avalanche photodiode the action of light gives rise to a controlled avalanche breakdown. The cut-off frequency of the avalanche diode is in the gigaHertz range (the amplification bandwidth product can reach 80 GHz). As a result of the highly critical adjustment of the working voltage and other characteristics (dark current), switching with avalanche diodes is very costly. Schottky photodiodes are used in position sensing applications as they can be manufactured as large area diodes. They can be manufactured to measure lateral deviations of a light beam from an optical axis with a length of up to 20 cm. Circular or quadratic surfaces of up to 10 cm$^2$ are divided into four quadrants in order to detect any deviations from two optical axes. Each quadrant has a terminal, and a common backplate electrode is also provided. If a directed beam hits the centre of such an arrangement no electrical signal is generated. If the beam should deviate from the centre, two signals proportional to the deviation are generated ($x$ and $y$ direction). These diodes are used in automatic plotting systems [44].

### 3.7.3   Signal recognition in photoelectric scanning

The recognition of photodetector signals is simple so long as there is a considerable difference between the bright and dark signals. This primarily affects divisions with coarse screening. In this connection, coarse screening refers to divisions up to a resolution of about 0.1 mm. However, the limit for the resolution of line scales is currently 0.1 $\mu$m. Because of this, special circuits are required for the photodetectors. By distinguishing only between the two states 'light' and 'no light' each photodetector signal has a d.c. component. In this case, both the bright current and the dark current are temperature-dependent and the great temperature dependency of the dark current in particular causes interference. Added to this, especially in the case of very fine divisions, is a greater constant light component as a percentage of the useful signal, which is due to the fact that the distance between the scale and the scanning reticule cannot be reduced as and when required. As only approximately sinusoidal signals are obtainable with fine divisions, the accuracy of measurement is very dependent on the trigger of the downstream pulse shaping stage. One way out is to use a reference photodiode to generate a d.c. signal which alters in the same way as the d.c. component of the scanning signal in the event of temperature fluctuations or ageing of the light source and is therefore suitable as a variable voltage for the trigger levels. It is more convenient to use two photodetectors to generate a logical signal. The two photodetectors are placed half a grid division apart and produce two modulated signals phase-shifted by 180°. As the d.c. component has the same polarity in both cases, a subtraction of the signals results in a doubling of the a.c. component amplitude at the same time as the d.c. component is dropping, Figure 3.61. Specially selected photodetectors should be used for this task. Switching amplifiers or comparators are used to convert the analogue signals of the photodetector into digital signals with clearly defined states. At a specific input voltage these amplifiers switch over very rapidly from one logical output state to another. If, however, the input voltage is low or the measurement process remains in one position in which the photodetectors emit a signal close to the switching wavelength, there is a danger that a switching amplifier will switch over many times. The switching over can also be caused by interference voltages overriding the signal voltage. In order to suppress any unwanted spike pulses released by it, the switching amplifier is designed for hysteresis. For this a certain mechanical movement of the system is required before switching of the trigger circuit occurs. Figure 3.62 shows a section of a grid scale and the associated analogue signal and associated digital signal. Point 1 may be considered the optimum forward break-over point whereas in fact, depending on the direction of motion, it is Point 2 or Point 3. In position measuring an error would be detected here and the greatest possible position deviation is defined as scanning error $\varepsilon_{max}$ [35].

### 3.7.4   *Measurement processes using static scanning reticules*

Gratings with alternating transparent and opaque sections modulate the amplitude of the light, and are therefore described as amplitude

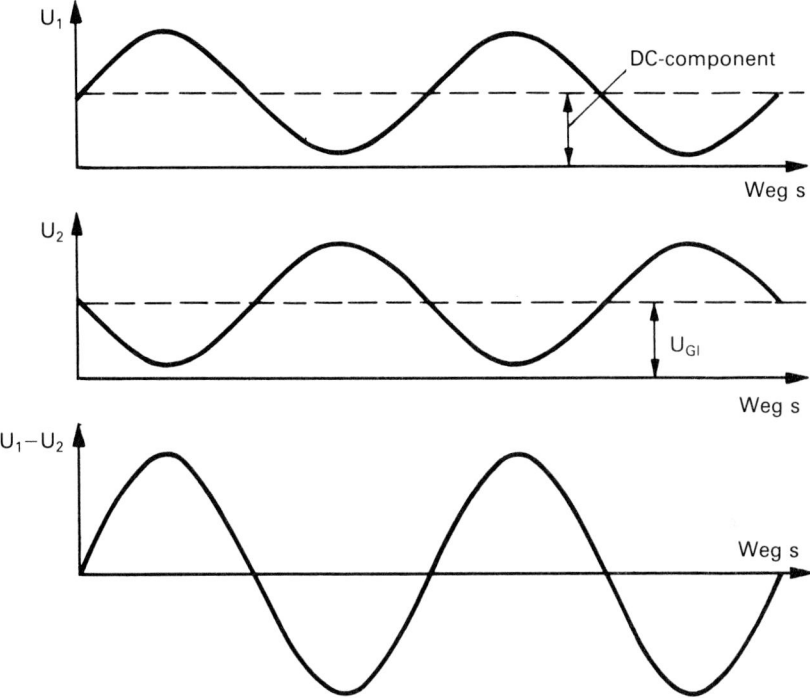

Figure 3.61 Suppression of the direct current component in photodetector signals. $u_{G1}$, Direct current component; $u_1$, signal from photodetector 1; $u_2$, signal from photodetector 2; $s$, displacement.

gratings. Unlike the transmitted light process, in the incident light process the grating division is in the form of sections of differing reflectivity on an opaque scale (usually made of steel). The principle of the transmitted light process is shown in Fig. 3.63. The glass scale bears a line graduation which may be produced by scribing, etching, vacuum evaporation of metal or photography. The scale is illuminated by a light source with condenser lens to produce parallel light. The scanning reticule which is used as a reference is located between the light source and the scale. In this case the photodetectors located on the opposite side of the scale are photovoltaic cells.

If the scale is moved relative to the scanning assembly, the illumination of the photodetectors fluctuates periodically, producing similarly fluctuating electrical signals. As a rule, each photodetector scans several grating periods simultaneously. In this way the illumination becomes sufficient, and at the same time averaging compensates for graduation errors and localized accumulations of dirt on

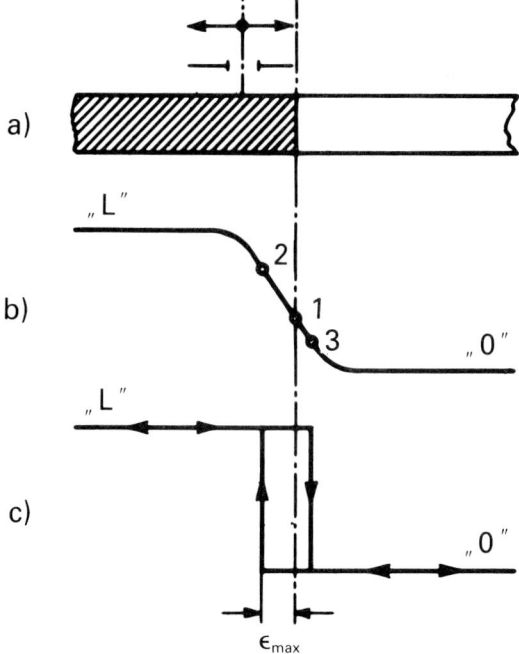

Figure 3.62   Definition of scanning uncertainty. $e_{max}$, Maximum scanning uncertainty; 1, optimum switching point, 2, 3, actual switching points.

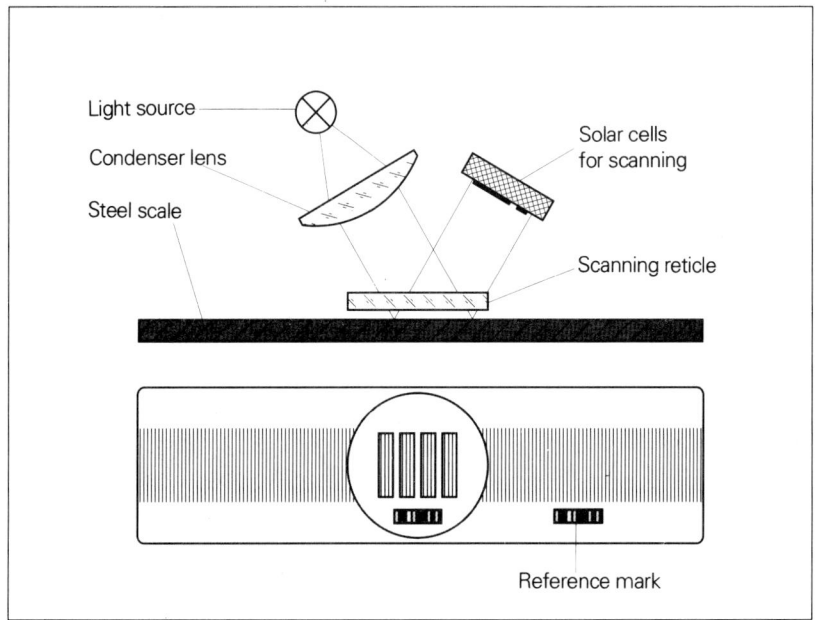

Figure 3.63   Principle of the direct light method.

Figure 3.64    Linear measuring system. (Picture courtesy of Heidenhain.)

the scale. Figure 3.64 shows an example of a high-resolution, high-precision linear measurement process which has a grating period of $10\,\mu$m, making possible measuring increments of $0.5\,\mu$m. In addition, the same manufacturer even offers a version in which the resolution is $0.1\,\mu$m. Both systems can be supplied to a class of accuracy of $\pm$ $\mu$m [45;46;47].

In the incident light process, the illumination and the photodetectors are on the same side of the scale. This offers considerable design advantages since, unlike the transmitted light process, the scale does not have to be clamped. Furthermore, the scale can be made of unbreakable material (not glass) with the same coefficient of expansion as most materials (e.g. steel, where $a = 12 \times 10^{-6}$ K$^{-1}$), Figure 3.65. Usually, however, glass scales with a thermal expansion coefficient of $10 \times 10^{-6}$ K$^{-1}$ are used, which is equivalent to that of steel or grey cast iron. Light from the light source falls through the scanning reticule, is reflected from the scale and again passes through the scanning reticule and hits the photodetector. When the scale is moved relative to the scanning head, brightness fluctuations are again produced on the photodetectors. Maximum illumination is attained when the transparent gaps in the grid pattern in the scanning head are over the reflective lines on the steel scale. The disadvantage

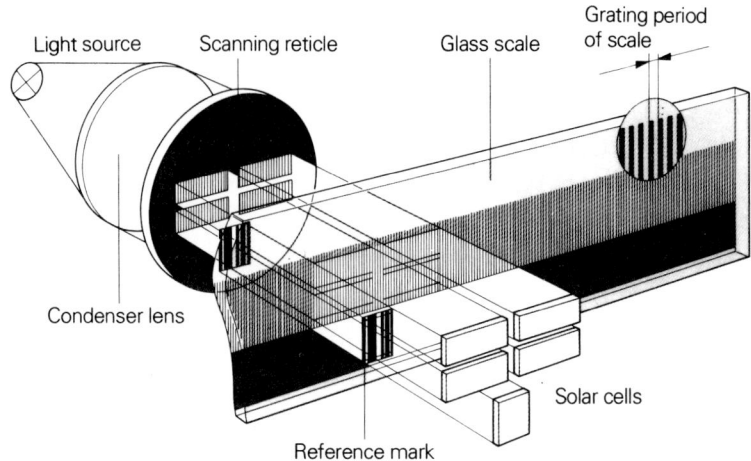

Figure 3.65   Principle of the incident light method.

of the incident light process in relation to the transmitted light process is that the relationship between the bright and dark sections is less satisfactory as a result of the absorptive or reflective action of the various grating sections, and additional loss of light occurs due to the light passing through the scanning reticule twice. At the highest class of accuracy of $\pm \mu$m with a grating division of $10 \mu$m, $0.1 \mu$m is given as the smallest measuring increment [45]. This applies to scales with amplitude gratings. In the case of phase gratings (see Chapter 3.7.6) classes of accuracy of $\pm \mu$m are possible with grating periods of $8 \mu$m and measuring increments of $\pm 0.2 \mu$m. With metal scales, however, measured lengths of up to 30 m are possible. The maximum path feed rate is a function of the resolution of the scale. It is 30 m min$^{-1}$ at a resolution of $0.1 \mu$m.

### 3.7.4.1   Using spatial frequency filters to generate static sinusoidal signals

In incremental length and angle measurement processes, resolutions of $0.1 \mu$m are possible only via the indirect route of generation sinusoidal signals, since the smallest grating divisions which can be manufactured at reasonable cost are in the region of $8 \mu$m. In optical gratings, the resolution is physically limited by the wavelength of the light. The grating period must in any case be greater than the wavelength of the light.

When a line grating is scanned using a scanning graticule the photoelectric current is a periodic function of the shift $v$ between the scale and the scanning graticule. It can be shown in the form:

$$i_1(v) = \sum_{n=0}^{\infty} a_{1n} \times \cos n\, Kv \tag{45}$$

$$i_2(v) = \sum_{n+0}^{\infty} a_{2n} \times \cos n \left( Kv - \frac{\pi}{2} \right) \tag{46}$$

where $K = 2\pi/T$, $T$ is the index grating period or lattice constant and $n = 0, 1, 2, \ldots$. By using optical methods, a situation can be created where all the higher order components disappear and only the fundamental wavelength remains. One sine and one cosine signal is thus obtained which makes it possible by interpolation to increase the resolution by a factor of 10, 20, 40 or 100. In order to generate signals free of harmonic waves, the electrical filtration shuts off, as the signal frequency is a function of the relative speed between the scanning reticule and the scale grating. Optical filtration makes use of the moiré effect. When the two gratings are twisted slightly against each other, moiré fringes are produced (Figure 3.49) whose brightness range matches the brightness range obtained using parallel grating lines. The photoelectric signals produced by the movement still, however, contain harmonic wave components. Figure 3.66 shows the principle by which sinusoidal signals are obtained using a spatial frequency filter. The line scale (3) is illuminated by means of parallel light passing through a light guide (1) and a lens (2). The light guides prevent the scale being heated. If an even wave front, i.e. a parallel light bundle, hits the grating, some of the light is diffracted into individual orders of diffraction. A second diffraction of light occurs at the scanning graticule, followed by interference between light bundles running parallel with each other.

The interference produces the desired brightness range as a function of place (spatial frequency). When beams pass diagonally through the grating system, the optical path length of the individual orders of diffraction and therefore the phase position of the beams reaching the interference are different. The signals generated are not yet free of harmonic waves. The harmonic waves can be filtered out by means of an additional screen (scanning gate), by letting the scanning gate integrate the light over a specified plane. If, for

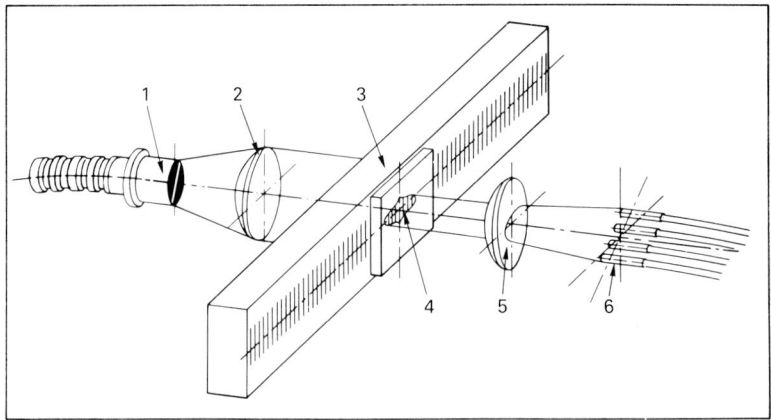

Figure 3.66   Principle of the PHOCOSIN method (Zeiss).
1, Light guide; 2, condenser; 3, index grating; 4, scanning grating; 5, light collector; 6, photodetector.

example, bright and dark points to be allocated to a harmonic wave are in this plane, then this harmonic wave will drop out. If a sinusoidal contour is selected for the scanning gate, all the harmonic waves will be filtered out.

In order to obtain purely sinusoidal signals the Fourier coefficients $a_{10}$ and $a_{20}$, i.e. the d.c. component, must be set to zero. In addition, two signals phase-shifted by 180° must be generated and subtracted from each other. A total of four 90° phase-shifted signals are required in order to achieve freedom from harmonic waves but to be able to carry out directional recognition. These four signals are generated by the photodetector shown in Figure 3.66. An accumulation of dirt on the grating, graduation error, eccentricity error (in angle encoders), etc. can give rise to different amplitudes of the signals which have to be subtracted in pairs when the scanning points for the four signals required are adjacent to each other. This source of error can be avoided by use of a single scannable field. This scannable field will be at a specific distance from the grating scale.

In the case of photoelectric angle encoders, the scaled discs may be made of metal, glass, plexiglass or film material, depending on the intended application. Etched or galvanically deposited metal plates are especially suitable for applications involving a low pulse rate. They have the advantage that the transparent sections are absolutely transparent and the lines are absolutely opaque. Glass discs are used

Figure 3.67   Shaft encoder with solid shaft (left) and hollow shaft (right).
(Picture courtesy of Erwin Halstrup Multur GmbH.)

in high-resolution encoders. The grating lines are applied either by
etching, vacuum metalizing, galvanizing or photography. Glass is
stable when subjected to temperature fluctuations and is insensitive
to moisture.

Because it is more readily machinable and lightweight, plexiglass is
used in medium resolution encoders.

Film material with a thickness of about 0.22 mm is used in many
applications where insensitivity to shock and vibration is required
(very light weight). Besides, it is easily workable. Figure 3.67 shows an
encoder for use on processing and finishing machines with a resolution
of up to 5040 positions per rotation. This encoder can be fitted with
either a solid shaft or a hollow shaft. The hollow shaft enables it to be
simply and compactly mounted directly on to a motor or gear shaft. It
is suitable for operation under severe shock and vibration stresses and
can therefore be directly mounted on presses and stamping machines.
Its maximum speed is 6000 r.p.m. It emits three rectangular output
signals: 2 pulse trains A and B phase-shifted by 90° and a reference
signal which encloses a positive half-wave. Another interesting feature
is the fact that the supply voltage may range anywhere between 10 and
30 V. The encoder shown in Figure 3.68 is designed for applications

Figure 3.68   Incremental shaft encoder for a resolution of 0.0001°.
(Picture courtesy of Heidenhain.)

requiring an extremely high resolution and accuracy. A minimum measuring increment of 0.0001° is possible with 36000 lines. This gives a 25-times graduation and a 4-times analysis. The output signals are two approximately sinusoidal signals with a 90° phase shift and an additional reference signal.

### 3.7.5   *Dynamic photoelectric grid scanning*

Unlike the static scanning of grid scales which produces d.c. signals when at a standstill, in dynamic scanning an a.c. signal is also generated when at a standstill. This creates the possibility of interpolating the encoder signals electrically and checking the scale automatically for dirt.

The scale consists of a steel support on to which a glass scale has been fixed so that the grid is facing the support. The grid consists of alternate reflective and non-reflective sections. As the glass is harder than steel it is not possible for the grid pattern to be scratched by swarf. The grating period is 0.635 mm = 1/40 inch. The interface between two guide pieces can therefore be relatively wide, which means that any desired measured length can be achieved without difficulty by lining up sections of the scale.

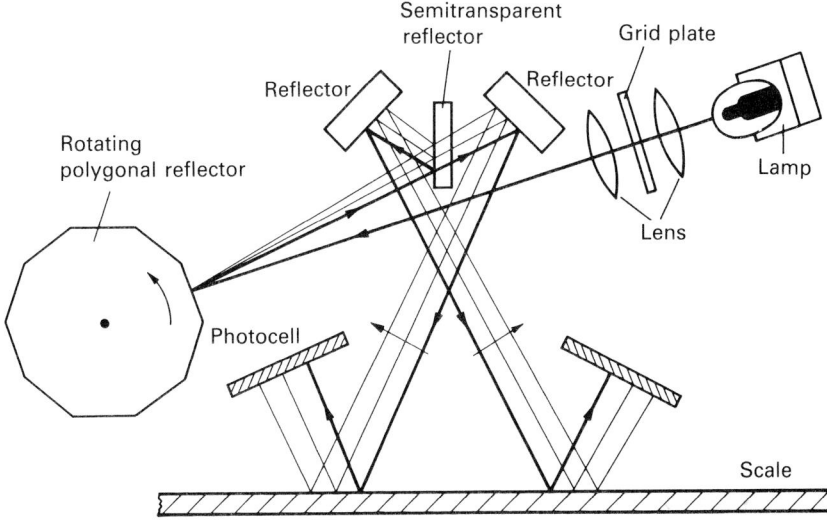

Figure 3.69   Ray path during dynamic scanning of a grid scale.

The classic method of dynamic scanning is to use a rotating polygonal reflector, Figure 3.69. This has 262 high-precision facets and rotates at a constant speed of 375 r.p.m. The principle of dynamic scanning is that the image of a grating is moved over the scale, the movement being caused by the polygonal reflector. The scale and the image of the grid have the same graduation. If the grid image projected falls on to the non-reflecting section of the line grid, no light is reflected. If it hits the reflection grating lines, a maximum amount of light is reflected. A photodetector which converts the reflected light into an electrical signal, produces an a.c. voltage, the phase of which contains information on the relative position of the sensing head and the scale. In order to analyse this information, a reference quantity is required, which is obtained by using a semi-transparent reflector to split a partial beam from the beam reflected by the polygonal reflector. This partial beam is reflected twice before it falls on to the scale, which it scans from left to right, as shown in Figure 3.66. The beam allowed through by the splitter is reflected only once and moves from right to left over the scale. In this way two a.c. voltages are obtained with a phase difference proportional to the position of the sensing head [49].

In a further develoμ nent of this principle of measurement, the mechanical movement of the polygonal reflector is replaced by an

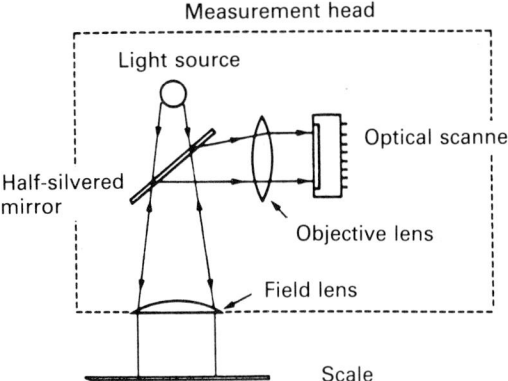

Figure 3.70   Mode of operation of a dynamic measuring head for a linear measuring system.

apparent movement of the photodetector [50]. Instead of a single photodetector, an array of 220 photodiodes arranged in a line is used. An image of the scale is projected through a lens on to this line, Figure 3.70. The scanning of this image takes place as in a television camera, with the difference that only one line is scanned back and forth periodically.

During this time the individual photodiodes are switched on and off one after the other. Where a reflective section is projected, a higher photoelectric current flows than with a non-reflective section. In an industrial measurement process (Figure 3.72) the semi-transparent reflector is replaced by a highly reflective mirror with a geometric graduation in the beam path, as shown in the diagram in Figure 3.70. In this way, loss of light due to the semi-transparent reflector, where in practice at least three-quarters of the luminous energy can be lost, is avoided. Although the beam path inclines about an angle of 3° to the scale, changes in the distance between the sensing head and the scale have no effect on displacement measurement. The effective scan line of the sensing head continues to be at right angles to the grating lines in the scale. The beam path between the sensing head and the scale is distinguished by a telecentric image. The light source projects an image through the field lenses into the optical centre of the subject, the field lenses acting as condenser. This removes problems arising from the distance between the sensing head and the measuring tape, as although the sharpness of the image changes, the graduation does not. The optical instrument, which consists of field lenses and an objective, produces a sharp reduced image of the scale grid on the light-sensitive area of the sensor, which is housed in a 16 pole case with a glass window. The actual photoelectric sensor consists of a line of 220 cells spaced at an average distance of $20\,\mu$m apart and each individual cell is $10\,\mu$m wide and 1.8 mm long. The total active area is $1.8 \times 4.4$ mm. The scannable field contains 22 grid divisions with 10 cells per division. Ten photocells with a total length of $200\,\mu$m are equal to one grid division of $635\,\mu$m. The reduction ratio produced is therefore $635 : 200 = 3.175$. This gives the width of the imaged grid, where $3.175 \times 1.8$ mm $= 5.75$ mm. The method of simulating a movable grid is to divide the line of photodiodes into 22 groups of 10 photodiodes each and activate five which are adjacent to each other. These groups of five are shifted incrementally from home position to the left or right by electronic means. Figure 3.71 shows the overlaying of the moving grid with the image of the scale.

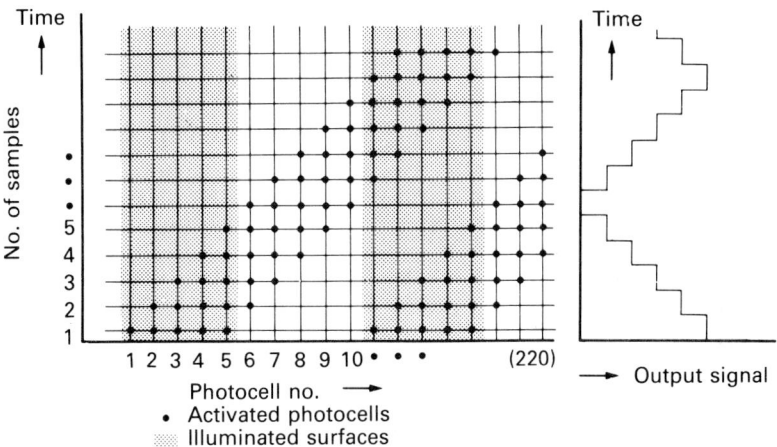

Figure 3.71  Electronic simulation of a movable measuring grid. Shown here is the movable measuring grid combined with the image of the scale. The output signal *MS* is approximately sinusoidal in shape.

The output of the scan lines, the measuring signal *MS*, resembles a flattened sinusoidal shape which contains the relative position data. Each of the 220 photocells must be connected to the output in a specific sequence in order to generate the correct *MS* signal. As cells 1, 11, 21, etc. are connected internally in exactly the same way as cells 2, 12, 22 and 3, 13, 23, etc., the problem is reduced to the switching of groups 1–10. This is simply achieved by means of a shift register, Figure 3.73. Each timing pulse shifts the contents of the shift register

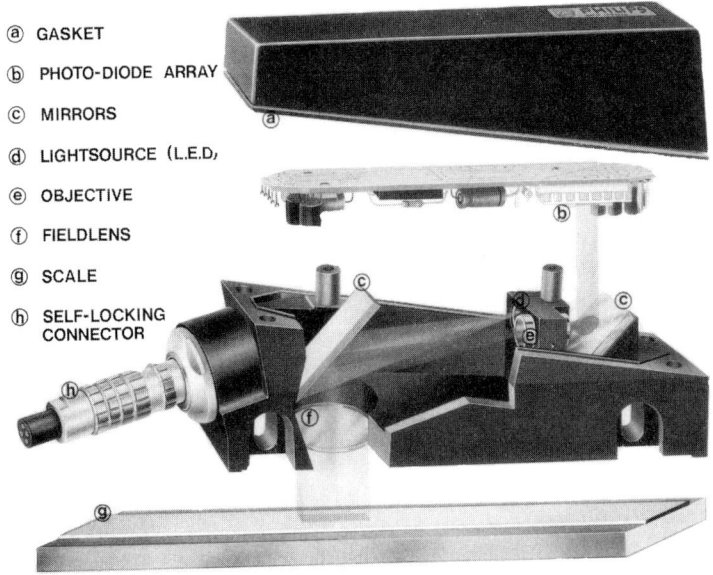

ⓐ GASKET

ⓑ PHOTO-DIODE ARRAY

ⓒ MIRRORS

ⓓ LIGHTSOURCE (L.E.D.)

ⓔ OBJECTIVE

ⓕ FIELDLENS

ⓖ SCALE

ⓗ SELF-LOCKING
   CONNECTOR

Figure 3.72   Construction of a compact optical measuring head for the dynamic scanning of a grid scale. (Picture courtesy of Philips.)

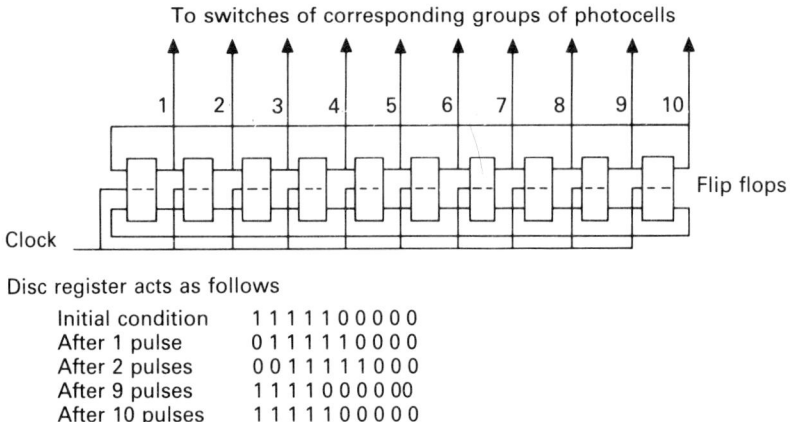

To switches of corresponding groups of photocells

1   2   3   4   5   6   7   8   9   10

Flip flops

Clock

Disc register acts as follows

| Initial condition | 1 1 1 1 1 0 0 0 0 0 |
|---|---|
| After 1 pulse | 0 1 1 1 1 1 0 0 0 0 |
| After 2 pulses | 0 0 1 1 1 1 1 0 0 0 |
| After 9 pulses | 1 1 1 1 0 0 0 0 0 0 |
| After 10 pulses | 1 1 1 1 1 0 0 0 0 0 |

Figure 3.73   The circuit required for simulating a moving grid has been designed as a 10 bit shift register.

to the right while the output is fed back to the input. The output of each flip-flop circuit activates the corresponding group of cells. The *MS* output is still not exactly sinusoidal, but has the shape of a step-type voltage. The sinusoidal signal is obtained by filtering the ground wave and then fed to an interpolation unit. In order to obtain the desired resolution of $0.5\,\mu m$ the period of the *MS* signal, corresponding to 0.635 mm, must be divided 1270 times. The way this happens is that the timing frequency of the shift register of 27.5 kHz is obtained from an oscillator frequency of 2.5 MHz by dividing it by 127. As the *MS* signal, owing to the 10 consecutively scanned photodiodes, has a frequency, lower by a factor of 10, of 2.75 kHz, the period of the *MS* signal can be simply interpolated when the measurement process is at a standstill using the oscillator frequency of 3.5 MHz. In this way there are 1270 oscillator pulses between every two *MS* pulses. If the sensing head moves to the left in relation to the scale, the frequency of the measuring signal *MS* increases. As a result, fewer than 1270 pulses are generated between each two *MS* pulses. Where there is relative motion between measuring head and grid scale with angular velocity $v$ the relation

$$MS = A \sin\left(\omega t + 2\pi \frac{x}{T}\right) \tag{47}$$

applies to the measuring signal *MS*. *A* is the amplitude, *T* the scale division and $x$ the position of the sensing head in relation to the scale; $\omega$ increases or decreases depending on the direction of motion, because the image of the scale on the scan line and the scan move relative to each other. A movement to the right results in a lower *MS* frequency and therefore to more than 1270 pulses between two *MS* pulses. These difference pulses are emitted as position-counting pulses, so that the difference symbol is used to indicate the direction of movement. If the absolute position of the sensing head is essential, an accurate zero mark must be used. This function may be fulfilled by adding a relatively inaccurate actuator. A reference signal *RS* is required for this. The *RS* signal is produced by dividing the timing pulse signal by 10:

$$RS = B \times \sin(\omega t + \varphi) \tag{48}$$

When the sensing head moves along the scale, each scale division of $635\,\mu m$ (e.g. through the point in the phase balance between the signals *RS* and *MS*) can be marked. If the corresponding division is selected using a rather rough and ready switch unit, e.g. microswitch, and the two signals are linked, an accurate reference point signal is produced. The linearity of the process is very dependent on the number of photosensors used, as the steps in the *MS* signal become fewer as the number increases. A further influencing variable is the quantity of the area imaged by the scale. The greater the area of scale grid scanned, the smaller any influence of scale irregularities, dirt on the grid, etc. A considerable advantage of a dynamic system, i.e. a system which produces an a.c. signal when at a standstill, is that it is possible to filter out the sinusoidal ground wave from the measuring signal. This can be interpolated with a high degree of accuracy. An interpolation factor of 1270, as in the present case, is not possible in static systems.

### 3.7.6   Length and angle encoders with phase grating

The development and production of semiconductor memories (the production of 64 Mbit memories is planned for 1996) requires measurement processes with resolutions in the nanometre region. A typical line width for a 64 Mbit memory is 0.3 $\mu$m (300 nm). Length measurement processes must have resolution in the nanometre region in order to ensure faultless control of the application of these configurations. The high precision finishing of complex reflective surfaces as required for directing beams in laser cutting also need resolutions below 20 nm.

The grating divisions with alternating transparent and opaque sections as used hitherto are unsuitable as they make impracticable demands on mounting accuracy. Although laser interferometers are suitable in principle, they require a disproportionate amount of equipment as a result of their dependency on the refractive index of air (and therefore the light wavelength), on temperature, air pressure and the exact gas components [51].

This task of carrying out length measurements in the nanometre region can still, however, be completed using scales with phase gratings where diffraction effects are used to obtain signals.

The gratings with alternating transparent and opaque sections considered up to now give rise to a modulation of light amplitude and are therefore known as amplitude gratings. In phase gratings, however, it is the phase position of the light waves which is modulated. Regularity is used so that the rate of diffusion of the light is inversely proportional to the refractive index $n$ of the medium through which the light is diffused. The refractive index $n$ of glass is between 1.45 and 1.75 and that of air is 1. If a parallel bundle of rays is sent through a glass plate with a stepped surface, as shown in Figure 3.74, the parts of the bundle with a longer path in the glass undergo a phase delay in relation to the parts of the bundle which first emerge from the glass. It is advisable to select the difference in thickness $d$ between the steps in the grating in such a way that the

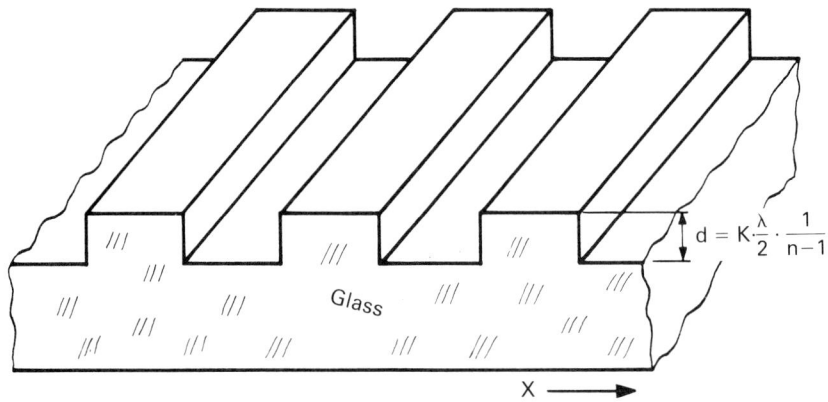

Figure 3.74   Principle of the phase grating.

path difference is $g/2$. The emerging parts of the bundle then interfere, producing a diffraction pattern as shown in Figure 3.75. In this
example, only the first, third and fifth orders of diffraction have been
shown, as these account for most of the energy. The ±1st orders
contain 40%, the ±3rd orders 4.4% and the ±5th orders 1.6% of the
transmitted energy. If only the ±1st orders are analysed, as is the
case in this example, then at least 80% of the light energy is used for
position sensing. In contrast to this, an amplitude grating has
basically 50% light loss through the opaque sections at first. In addition, the amplitude grating produces a strong 0th order that it
accounts for only 12% of the energy. An amplitude grating performs
the task of splitting a bundle of rays into two coherent partial
bundles just as well, in principle, as a phase grating, but has the disadvantage of being much less efficient.

The reason for using only the ±1st orders is that they contain only
the fundamental of the information to be transmitted, the grid structure. This can be explained in greater detail.

The geometrical shape of the grating describes information which
is transmitted by the light. In this way the light takes on the role of a
carrier frequency which is modulated with this information. The information itself can be separated into a harmonic series by means
of Fourier transformation. The Fourier series for a rectangular
sequence is:

$$y = \frac{4}{\pi}\left(\sin x + \frac{1}{3}\sin 3x + \frac{1}{5}\sin 5x \ldots\right) \qquad (49)$$

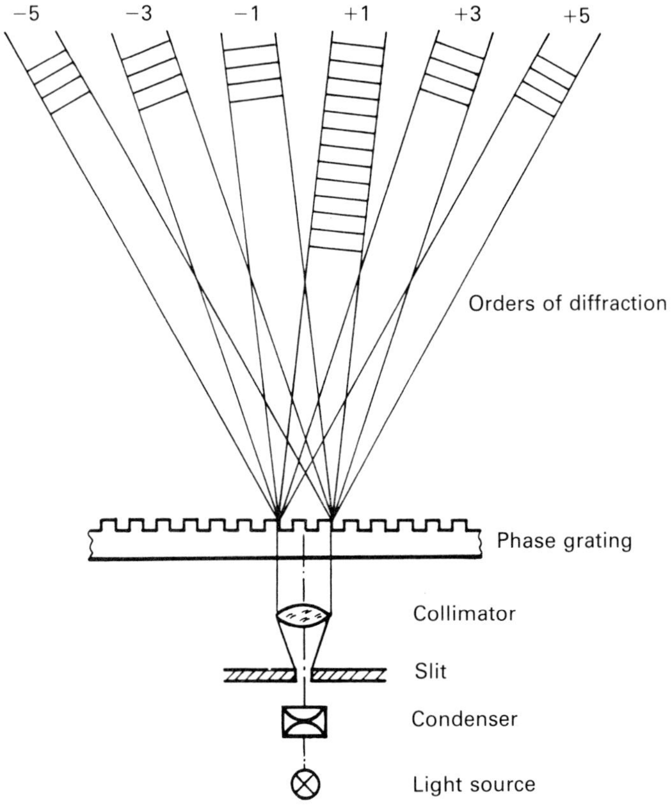

Figure 3.75   Diffraction at the phase grating.

i.e. only the odd harmonics are present. Here $x$ is the three-dimensional coordinate (Figure 3.74). The spectrum is referred to as the spatial-frequency spectrum of the grating. The orders of diffraction have the property that they each contain a spatial frequency of the same order and only that frequency. Thus the 1st order contains the fundamental of the spatial frequency, the 3rd order the 3rd harmonic, etc. When the phase grating is displaced in relation to the light slit, the phase positions of the light waves change in the various orders of diffraction. In the +1st order of diffraction, displacing the grid by one period results in a phase shift of the light wave of $+2\pi$, and in the −1st order of diffraction a phase shift of $-2\pi$. In the +3rd order, the same grid displacement results in a phase shift of $+6\pi$ and correspondingly the −3rd order has a phase shift of $-6\pi$. Basically,

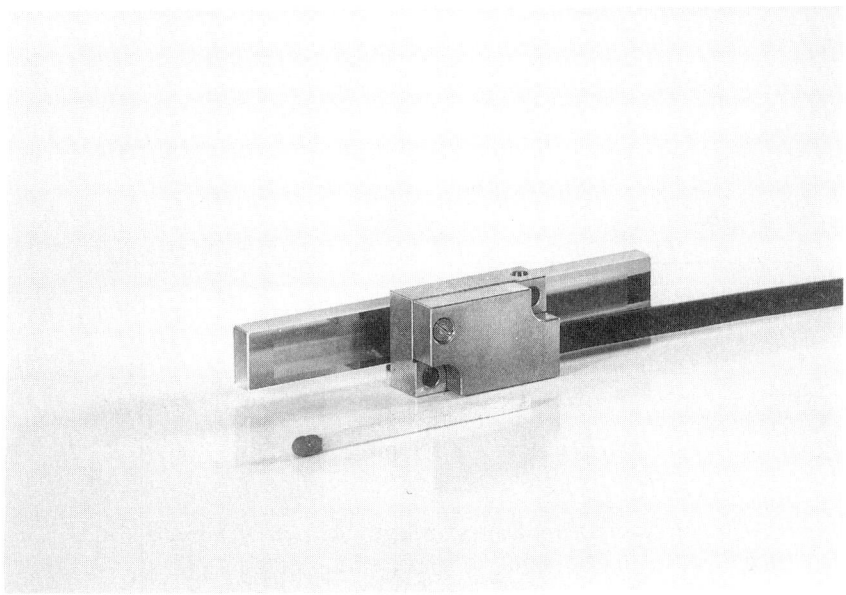

Figure 3.76    Measuring instrument with $5\,\mu$m increment. (Picture courtesy of Heidenhain.)

orders with the same ordinal number but a different preceding sign have opposite phase positions.

A length encoder with phase grating, Figure 3.76, uses these characteristics of diffraction in a phase grating [51;52;53]. This has a grating period of $4\,\mu$m which results in a signal period of $2\,\mu$m. A resolution of 5 nm is obtained by means of an additional electronic subdivision of the measuring increments by a factor of 400. A still higher resolution can be obtained by an encoder using the same principle but with a grating period of $1.6\,\mu$m. After optical quadrupling of the signals and subsequent electronic subdivision by a factor of 400, a resolution of 1 nm is reached [53]. The operating principle can be explained as follows. The scale (Figure 3.76) has a phase grating with a grating period $g$. The division ratio (line/space) is $1:1$. As already shown in Figure 3.75, the difference in thickness between the grid increments is selected in such a way that the path difference between the emerging partial bundles of rays is $g/2$ and no 0th order of diffraction arises. If the grid is moved by the amount $u$, the phase of the diffracted wave of the 1st order is changed by

$$\Omega = \frac{2\pi u}{g} \tag{50}$$

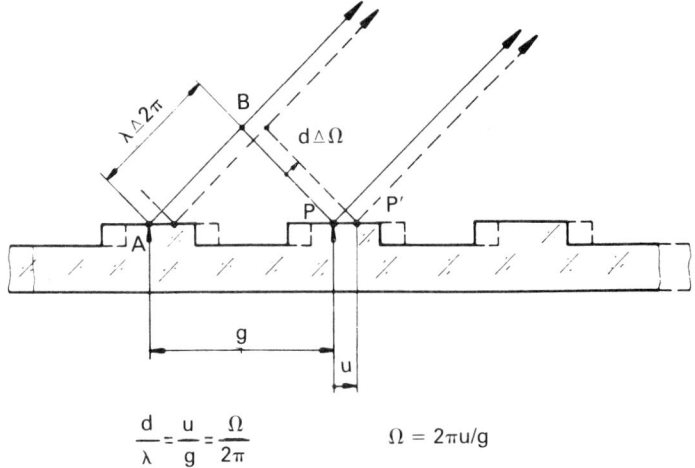

Figure 3.77   Phase shift of the 1st order of diffraction when a grid is displaced. u, Grid displacement; $\Omega$, phase shift.

This can be visualized from the geometrical relations (Figure 3.77): in the output position of the grid, the point $P$ is the place where the diffracted wave of the 1st order has the phase position $\Omega = 0$. If the grid is displaced by an amount $u$, then the point $P$ drifts to position $P'$. The wave emerging here has in addition covered the path difference $d$ in relation to the original wave emerging from $P$. The requirement for the diffraction in the +1st order produces for a beam emerging from point $A$ ($A$ being a grating period between $g$ and $P$) the path difference $A - B = g$ corresponding to a phase difference of 2 $u$. Geometric similarities then arise for the phase:

$$\frac{\Omega}{2\pi} = \frac{u}{g} \tag{51}$$

and

$$\Omega = \frac{2\pi u}{g} \tag{52}$$

A phase shift of a similar size is produced for the −1st order of diffraction but with the opposite preceding sign.

The structural design of a length measurement process with a phase grating is shown in Figure 3.78 and the associated optical diagram in Figure 3.79 [51]. A light source (1) (LED) emits light through a condenser lens which produces an even illumination wave, a first phase grating known as the index grating. This index grating effects a

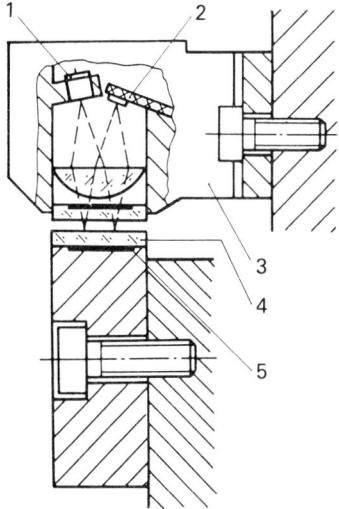

Figure 3.78   Open linear measuring system. 1, Light source (LED); 2, photocells; 3, scanning head; 4, protecting lens; 5, phase grating scale.

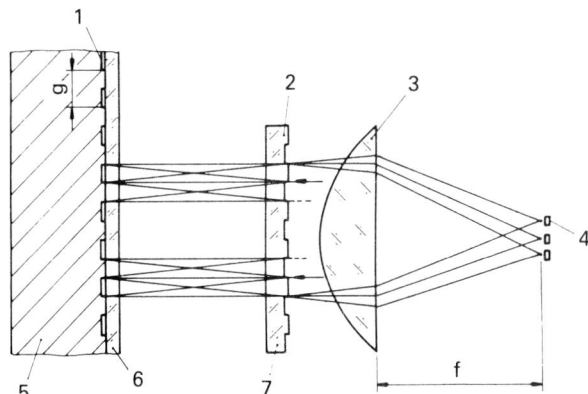

Figure 3.79   Optical diagram of interferential linear measuring system, scale and scanning plate (index) with phase grating division. 1, Phase grating; 2, phase grating; 3, condenser; 4, photocells; 5, steel scale; 6, cover slip; 7, scanning plate.

first diffraction, in three main directions. These three waves are reflected at the actual scale (5) and again diffracted. The scale is made of reflective steel on to which have been applied, at a grating period of 8 $\mu$m, rectangular increments in gold 4 $\mu$m wide and about 0.2 $\mu$m thick. The reflected beams pass for a second time through the index grating and are again diffracted. The lens then produces three images

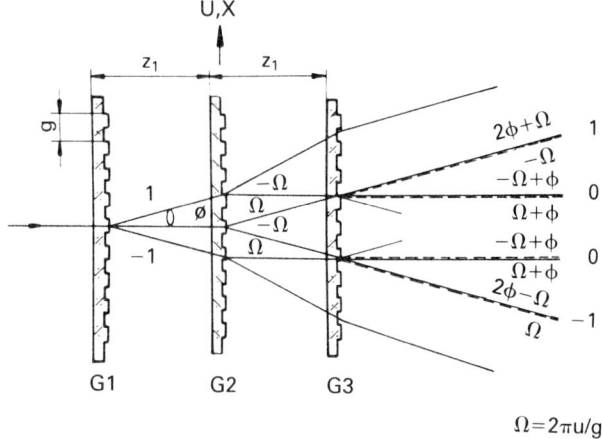

Figure 3.80   Operation of the interferential measuring system.

of the light source which are converted by three photocells (2) into electrical signals. The unit includes a three-grid interferometer, as previously described by Willhelm [64]. The principle of the three-grid interferometer can be seen in Figure 3.80. Only the three grids are considered here, the other components such as the light source, photodetector and condenser having been disregarded. To simplify matters the final glass thickness is disregarded and it is assumed that the incoming beam has only one even wave of wavelength $\lambda$ and strikes the three grids vertically. In the first grid G1 (corresponding to the first passage through the index grating Figure 3.80) three main orders of diffraction are produced (−1; 0; 1). The index grating is designed so that the 0th order lags the ±1st orders by the angle $\phi$. The three beams then strike the actual scale (G2), which is designed so that no 0th order arises. This means that three partial beams are only diffracted into one +1st order and one −1st order. When the scale is displaced in relation to grids G1 and G3, which match the index grating, the beams in the +1st order are phase shifted by the angle $+\Omega$ and in the −1st order by the angle $-\Omega$. As it passes through grid G3 (or if passing for the second time, through the index grating) each of the six partial beams is again diffracted into a −1st, 0th, and +1st order and the three waves of the same order are superposed (interfere). The phase position of the newly produced wave after the superposition is derived from the sum of the phase shifts which the individual wave orders have undergone at the three grids. It can be shown that if illumination is provided by a non-monochromatic and spatially incoherent light source only the +1st, 0th and −1st order contribute to the production of a signal. The result obtained (for deduction see [51]) for the intensities of the three resulting waves with the −1st, 0th, +1st directions on the three photodetectors is:

$$I_0 = 2\left[1 + \cos\left(\frac{2.2\pi \times u}{g}\right)\right] \qquad (54)$$

$$I_{+1} = 2\left[1 + \cos\left(\frac{2.2\pi \times u}{g} + 2\varphi\right)\right] \qquad (55)$$

The three photodetectors (photocells) emit three signals, two of which are phase shifted by +2φ and −2φ, respectively, in relation to the third signal. All three have a period which matches the grating period of the phase grating.

In practice the phase grating is now designed so that 2φ = 120°. This means that three electrical signals are obtained which are the equivalent of a rotary current system from which the two phase-shifted signals required for further signal processing can easily be obtained.

A high-resolution angle encoder with phase grating is shown in Figure 3.81. A G$_a$A$_s$ diode (a) is used as a light source, with a condenser (b) to produce an even illumination wave plus a slit diaphragm and a collimator to image the slit on the phase grating of the circular graduation.

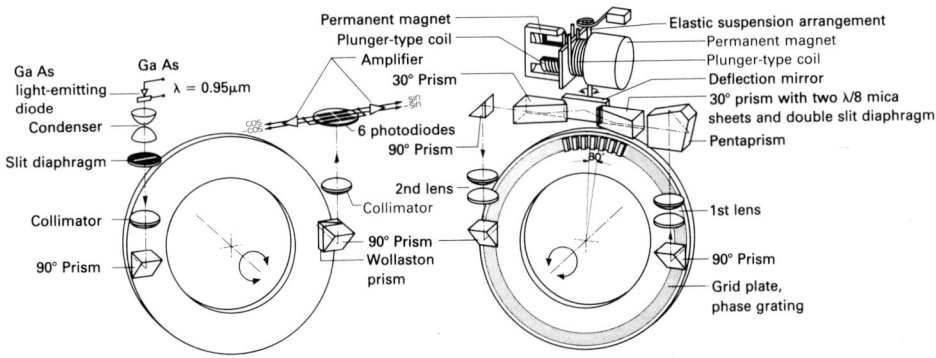

Figure 3.81   Angle encoder—built-in unit.

As shown in Figure 3.81, the difference in thickness between the grating steps must be such that no 0th order of diffraction is produced and 80% of the energy goes to the ±1st orders, the course of which is decisive for further observations. Two pairs of 90° prisms (e) are used to deflect the beam path in front of and behind the grid plate. After the first passage through the grid plate, it is mainly the ± 1st, the ±3rd and the ±5th orders of diffraction which are present. These are converted into parallel bundles through a first objective (r) and deflected again through the pentaprism (q). Following the pentaprism there is a 30° prism (p) on the leading surface of which slit diaphragms and two $g/8$ mica plates are located. The slit diaphragms admit only the ±1st orders of diffraction. These apertures act as spatial frequency filters, as only the fundamental of the image formation is admitted. During the subsequent interference of the two bundles of light, a purely sinusoidal wave of light as a function of the angle of rotation of the grid and a purely sinusoidal photodetector signal are obtained. This is of great importance for the multiplication of electrical signals. A characteristic of mica is that it is birefringent. This means that the light passing through it is polarized into two directions perpendicular to each other. In addition, a path difference occurs between the two directions of polarization as a function of the thickness of the mica plate. In the case in question, one mica plate has dimensions such that this path difference is $+g/8$ and the second so that it is $-g/8$.

Thus two polarized bundles perpendicular to each other are obtained, the modulation of which is in phase quadrature. The two bundles can be further processed independently of each other.

Firstly, however, the rest of the beam path should be described. The two 90° prisms (p, i), the deflecting reflector (o) and the moving coil system are part of an additional piece of equipment which is described in the next section. It is not required for the actual principle of measurement.

The beam path is deflected by two further 90° prisms, between which a second objective (1) is located, once more through the phase grating plate. Here diffraction again takes place. The orders of diffraction which occur are shown in Figure 3.82.

In Figure 3.82 we see the main beams of the ±1st order as focused through the second objective. The +1st order is split into a new series of positive and negative orders of diffraction which are given the suffix a. The corresponding group produced by the −1st order is given the suffix b. As a result of the different angles of incidence of the two main beams the following orders are superposed behind the receiving grating:

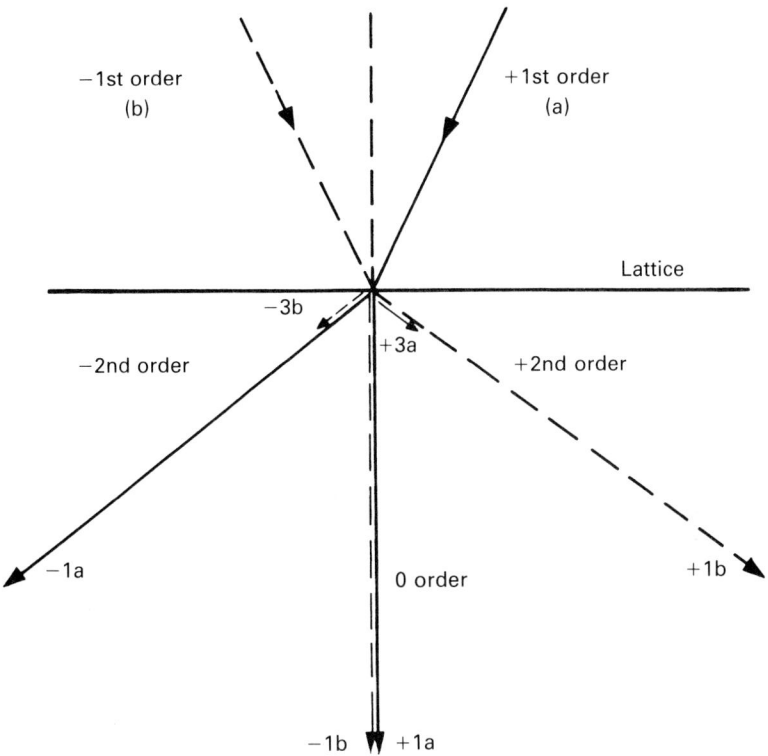

Figure 3.82 Superimposing the orders of diffraction emanating from the two primary ray bundles.

+3a on +1b, +1a on −1b and −1a on −3b.

The new orders arising from the superposition are then numbered +2, 0 and −2. As a result of the properties of the grating, +1a and −1a have 40% and +3a 4.4% of the intensity of the +1st order. The −1st order is proportioned accordingly. There thus exists between the interfering wave trains +1a and +3b and −1a and −3b a ratio of intensity of 9:1. This, however, plays only a subordinate role. As a result of square-law mixing in the photodetector, however, the interference contrast is still about 60% of the full modulation.

For further consideration it must be borne in mind that the second passage through the phase grating plate involves two bundles polarized normal to each other which are independent of each other. If the two bundles are allowed to interfere, as happens during the second passage through the grating, only the components of the same direction of polarization will interfere. In this way two completely independent interference processes are completed. These two processes can be separated by a beam splitter governed by the direction of polarization. A Wollaston prism is just such a splitter, dividing the incident light as a function of the direction of polarization into two beams which form an angle with each other.

In this way two separate images are obtained, the modulation of which is 90° phase-shifted as a result of the mica plates and therefore produces two 90° phase-shifted photoelectric signals. This means that the two signals required for deciding on direction are produced. However, it is not just two 90° phase-shifted signals which are produced, but in addition a further signal phase-shifted by 180° in relation to them. These enable the d.c. components in the photoelectric signals to be suppressed. The production of push–pull signals is described in Figure 3.82 and the preceding discussion: the phase difference between the new ±2nd orders is 180° in relation to the new 0th order. The photodetector can therefore be connected in parallel for the ±2nd orders of diffraction. They are back-to-back with the signals produced by the 0th order.

To complete the picture, the additional equipment mentioned previously, the core of which is a moving coil system, should now be described in detail. This is an apparatus for compensating index errors in the measurement base by means of an electromechanical tracking system. The major component of the tracking system is a moving coil system used to rotate a prism or a pair of prisms and a reflector (in an angle encoder). By rotating the prism in a length encoder, an additional path difference is introduced between the two sections of a bundle, which has exactly the same effect as moving the image or the grating. The sample applies to angle encoders. A tracking system is used to adjust the coil current so that the output signals of the two scanner amplifiers are of equal size, Figure 3.83 [54;55]. The two sinusoidal signals of the encoder are designated as sine and cosine as they are 90° phase-shifted relative to one another. The load-independent current in the moving coil system results in a proportional image displacement. Correction signals, capable of eliminating, for example, temperature influences or scale errors, can be fed in through an additional summing amplifier before the voltage–current converter. As soon as the two output signals from the encoder are the same size the count of the digital position value is in the micrometre region. The resolution of the length encoder is $0.1\,\mu m$ and is therefore directly comparable with that of a laser interferometer. The angle encoder has a resolution of 0.5 seconds of arc, which corresponds to 2 598 000 positions per rotation.

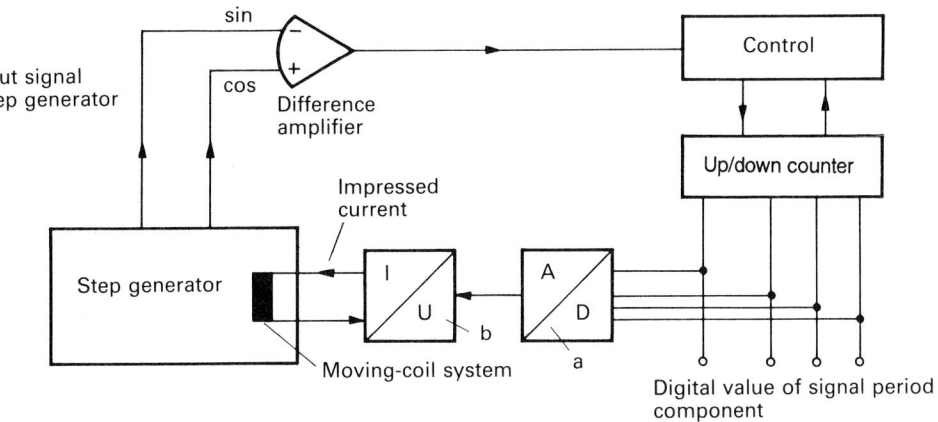

Figure 3.83  Tracking system of a precision measuring system. a, Digital-to-analogue converter; b, voltage-to-current converter.

# 3.8  Systems using a digitally absolute measurement base

Digital position sensing systems with a coded measurement base are analogue-to-digital converters which convert the analogue quantities of displacement or angle into coded electrical signals. As every length or angle value is allocated an unambiguous code word, these systems are referred to as absolute measurement processes [56;57]. In this case, the measured value is indicated solely by the coding or measurement base at the sensing point, irrespective of what has gone before, such as breakdowns due to power failure. In length measurement processes, a linear encoder is used as a measurement base and in angle measurement processes, a disc encoder. The physical possibilities for representing the code and the scanning methods are equivalent to those used to produce the screens in incremental systems. Thus the photoelectric scanning of optical screen traces is also prevalent in coded systems. What is therefore most interesting is the choice of code for the measurement base and the methods of scanning coded scales.

## 3.8.1  Coding

In this case, coding from an information source means quantizing and representing the information in numerical form, at the same time as reducing redundancy and information.

A code is an absolutely reversible, unique arrangement of two sets of characters. And a character may consist of a series of symbols. In the Morse alphabet, for example, each letter is allocated a specific series of dots and dashes: A·−, B−···, C−·−, etc. In another example, an integer can be represented as a sum of powers:

$$Z = a_n \times B^n + a_{n-1} \times B^{n-1} + \ldots a_0 \times B^0 = \sum_{i=0}^{n} a_i \times B^i \qquad (56)$$

$B$ is the basis of the numeric system, $0 \ldots n$ are the exponents. $a_i$ is the numerical value of the number at the $i$ th place. It can be represented by a character or a figure depending on the numerical system (code) used. The base $B$ denotes the numerical system. Besides the decimal system in normal daily use, where the base 10 is used, the binary system (base 2) in particular, as well as the octal system (base 8) and the hexadecimal system (base 16) play a major part in data processing. All these numeric systems are codes.

The binary system possesses only the two characters 0 and 1. A figure $Z$ is defined by:

$$Z = \sum_{i=0}^{n} a_i \times 2^i \qquad (57)$$

Example

$$59_{(10)} = 1 \times 2^5 + 1 \times 2^4 + 1 \times 2^3 + 0 \times 2^2 + 1 \times 2^2 + 1 \times 2^0 = 111011_{(2)}$$
$$= (32 + 16 + 8 + 0 + 2 + 1)_{(10)}$$

The octal system has the characters 0, 1, 2, 3, 4, 5, 6 and 7.

$$Z = \sum_{i=0}^{n} a_i \times 8^i \qquad (58)$$

Example

$$59_{(10)} = 7 \times 8^1 + 3 \times 8^0 = 73_{(8)}$$
$$= (56 + 3)_{(10)}$$

In the hexadecimal system there are a total of 16 characters: 0, 1, 2, 3, 4, 5, 6, 7, 8, 9, A, B, C, D, E and F.

$$Z = \sum_{i=0}^{n} a_i \times 16^i \qquad (59)$$

Example

$$59_{(10)} = 3 \times 16^1 + B \times 16^0 = 3B_{(H)}$$
$$= (48 + 11)_{(10)}$$

A further very common code is the BCD code where each decimal place has a binary code (BCD = Binary Coded Decimal).
   Example:

$$59_{(10)} = (0 \times 2^3 + 1 \times 2^2 + 0 \times 2^1 + 1 \times 2^0) \times 10^1$$
$$+ (1 \times 2^3 + 0 \times 2^2 + 0 \times 2^1 + 1 \times 2^0) \times 10^0$$
$$= (0101 \qquad 1001)_{(BCD)}$$
$$= (5 \quad 9)_{(10)}$$

The advantages of the BCD code are the clear arrangement and constant length of its code words. A disadvantage is that it requires more expensive switching technology for computing operations than is the case with the simple binary code.
   A feature of all the codes mentioned up to now is that each of their individual positions has a value. This means that a significance $W$ can be given for each bit position in a code word, so that the allocation requirement (i.e. the code) between the numeral or figure on the one hand and the binary code word on the other is given by the following equation:

$$Z = \sum_{i=1}^{n} W_i \times S_i \qquad (60)$$

$S_i$ is the significance of the $i$th position. In the binary code $S_i$ can assume the two values 0 or 1. In many cases, the possibility of attaching a value is however less important than other features. A requirement often made on analogue-to-digital converters is to measure unit distance. In this case adjacent code words differ by one position only. In a unit distance code a reading error at one position will not have any significant effect on the result. A code which is free of or low in redundancy is advisable for the coding of a signal source of, for example, a code scale or a code disc in order to keep equipment costs down. In information transmission, redundancy is usually deliberately added to the channel encoder in order to safeguard the transmission path and is then discarded in the channel decoder once the transmission is complete. Figure 3.84 shows the information flow during coded position sensing. Source coding is carried out as soon as the measurement base, i.e. the code scale, has been created. Subsequent channel coding is determined by the scanning method and the method of preparing the signals for transmission. The transmission path from the measurement process to the data processor, i.e. a cable, or the transmission path of the information from the source to a transmission device within the measurement process can be defined as a channel. The recognition of the signals produced by the scanning

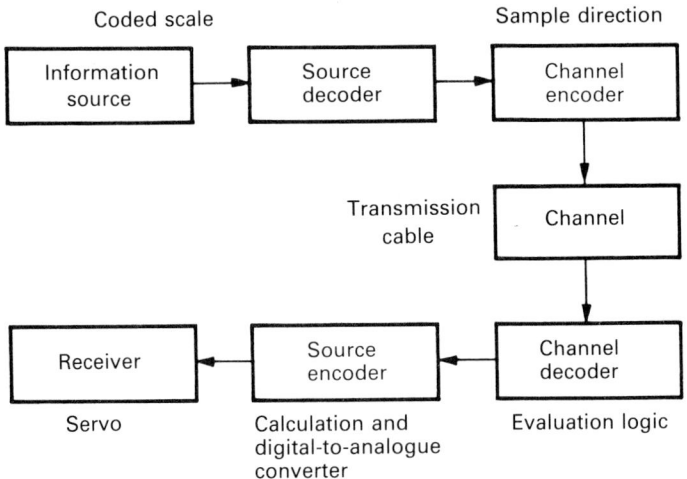

Figure 3.84   Information flow in coded position sensing.

electronics, irrespective of whether it happens in the measurement process itself or in the data processor, can be regarded as channel coding. In measurement processes using double scanning, V-logic takes on this function. The reconversion of digital information into analogue information must still be seen as source decoding.

The following considerations apply when selecting a code for a position sensing system.

- One consideration is which code the display, computing or control units to be connected to the measurement process are working in. This includes the question whether it is worthwhile changing the code (level coding).
- Security of data acquisition and transmission. The cost emphasis can be placed more on source coding or channel coding and decoding.
- Attaching values to the individual positions of the code. This includes the possibility of simple processing in arithmetic units.
- Formation rules and ease of recognition of the code.

The security of a code is dependent on the possibility of error detection and error correction. A measure of the detection or correction of errors is the number of positions at which a minimum of two words in a code differ [58;59]. This minimum number is described as the distance between the two words or the Hamming distance $d$. The number of positions occupied by one is described as the weighting of a code word. In many cases all the words are extended by adding a bit (check bit) to produce only odd or even code words, which therefore have an odd or even weighting. This results in a simple method of error detection for errors where only a single position is incorrect. This method is valuable as the statistical probability of single errors occurring is considerably greater than that of double errors or multiple errors. If the probability for a single error in a code word received is $p = 0.01$, then the probability for a double error is $p^2 = 0.0001$ and the probability for an error at $n$ positions of this word is $p^n$. The Hamming distance of a code with a check bit is $d = 2$ and it allows the detection of single errors. If three check bits are used it is possible to detect up to three errors in a code word or correct one single error.

The principle is that errors with the weighting $w$ are detectable when $w < d$. For error correction, it is assumed that $w < d/2$. Error correction must be visually presented in such a way that, if a single error occurs, a forbidden code word is produced which differs from the permissible code words next to it by only one bit. If three check bits are used, however, adjacent code words will differ by at least 4 bits. Statistical probability dictates that the most favourable conclusion is reached by replacing the erroneous code word with the permissible one which most resembles it. This method is also referred to as coding by maximum similarity. It follows from this, however, that a code has no absolute means of protection against errors. As adding further check bits does not increase the amount of actual information, this means that redundancy is added. Thus, although increased security of data transmission is provided, the cost of the transmitter, receiver and transmission channel is also increased at the same time. Many codes avoid code words which may all too readily occur in the event of malfunction. These are the zero-word (all positions occupied by zero) and the one-word (all positions occupied by one). An example of such a code is the three-excess code (also known as the Stiebitz code) in which the first and last three tetrads (groups of four) of the four-line binary code are defined as pseudotetrads, i.e. forbidden tetrads, Table 3.6.

The three-excess code has the additional advantage over the binary code and the BCD code that it allows a simple complement on 1 to be created in the binary system, which corresponds to the complement on nine in the decimal system.

Example   Decimal   Three-excess code
$$3 = 0110$$
(complement of nine)   $6 = 1001$ (complement of one)

Table 3.6  Binary codes

| Position values | Closed binary coding | Positional binary coding (BCD) | | | Closed binary coding |
|---|---|---|---|---|---|
| 8  4  2  1 | Binary Code | 8-4-2-1 Code | 3-Excess Code | Aiken Code | Gray Code |
| 0  0  0  0 | 0 | 0 |  | 0 | 0 |
| 0  0  0  1 | 1 | 1 |  | 1 | 1 |
| 0  0  1  0 | 2 | 2 |  | 2 | 3 |
| 0  0  1  1 | 3 | 3 | 0 | 3 | 2 |
| 0  1  0  0 | 4 | 4 | 1 | 4 | 7 |
| 0  1  0  1 | 5 | 5 | 2 |  | 6 |
| 0  1  1  0 | 6 | 6 | 3 |  | 4 |
| 0  1  1  1 | 7 | 7 | 4 |  | 5 |
| 1  0  0  0 | 8 | 8 | 5 |  | 15 |
| 1  0  0  1 | 9 | 9 | 6 |  | 14 |
| 1  0  1  0 | 10 |  | 7 |  | 12 |
| 1  0  1  1 | 11 |  | 8 | 5 | 13 |
| 1  1  0  0 | 12 |  | 9 | 6 | 8 |
| 1  1  0  1 | 13 |  |  | 7 | 9 |
| 1  1  1  0 | 14 |  |  | 8 | 11 |
| 1  1  1  1 | 15 |  |  | 9 | 10 |

Complements are created in the three-excess code by transposing 0 and 1. Subtraction operations can therefore be performed with ease.

Nowadays most arithmetical units and numerical controls operate either in the binary code or the BCD code. As decimal representation is the most favourable for communication between man and machine, the BCD code is preferred in input and output units. In addition, it is often used in small computers for which internal code conversion would be too expensive. Larger systems on the other hand operate in the binary code alone. These two codes are therefore recommended first and foremost for digital position sensing systems. One objection to their use, however, is the fact that neither is a unit-distance code. When passing from one number to the next, e.g. from 10111 (decimal 23) to 11000 (decimal 24), four positions change status at the same time, Figure 3.85. As a result of unavoidable manufacturing tolerances, the necessary simultaneity of this change cannot be achieved in all the positions affected. This means that incorrect numerical combinations may arise, at least for a short time.

There are three ways out of this difficulty [60;61].

- Avoid taking readings in the critical zone between two figures by blocking any reading in this area by adding a position and at the same time storing the preceding reading, Figure 3.86.

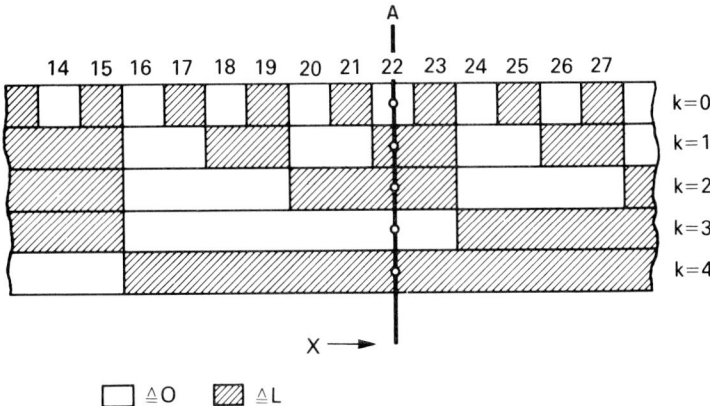

Figure 3.85 Five-track coded grid in binary code. *A*, Scanning line; *x*, direction of motion; *k*, order of code tracks.

- Use a unit-distance code, e.g. Gray code, Glixon code, O'Brien code, Tompkins code, Libaw–Craig code, Figure 3.87 [57;62].
- Use two scanning elements per position. From the scanning result of the finest position, which only one scanning element shows as the only track, it is decided which of the two signals of the subsequent position is to be analysed. This signal is again a criterion for the decision in the third position, etc.

The Gray code is the best known unit-distance code. It is, however, not so suitable for sub-coding decimal places, as passing from 9 to 0 is not a single step. The Glixon and O'Brien codes do not suffer from this disadvantage. The Tompkins code requires only two code tracks (with a total of four scanners) and the Libaw–Craig code just one code track, but with five scanners.

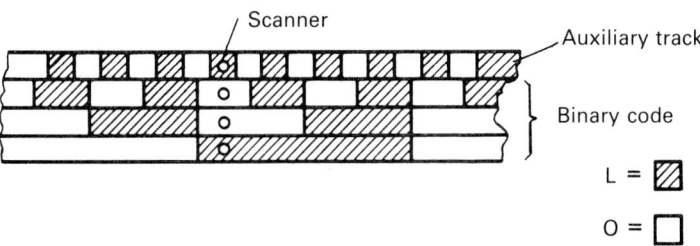

Figure 3.86 Code scale with auxiliary track. a, Scanner; b, auxiliary track; c, binary code.

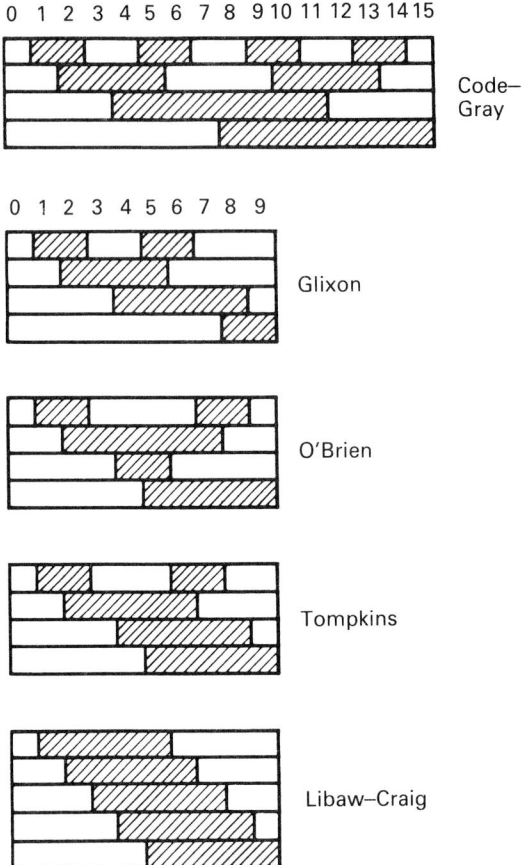

Figure 3.87   Unit-distance codes.

Gray numbers $G_n$ can be calculated from binary numbers $B_n$ using the following relation:

$$g_x = b_x \oplus b_{x+1}$$

where

$$G_n = (g_m, g_{m+1}, \ldots g_1, g_0$$
$$B_n = (b_m, b_{m-1}, \ldots b_1, b_0 \qquad (60)$$

It is possible to convert Gray code back into the binary code using a serial conversion circuit, as shown in Figure 3.88. The conversion of each individual bit is carried out in accordance with the relation:

$$b_x = b_{x+1} \oplus g_x \qquad (61)$$

In this case the bit with the highest value is the same in both cases. Each low value bit is a function of the higher value bit. This means that when using a logic circuit for the conversion, the conversion time is a function of the number of bits to be converted. Additional corrections are required when using excess codes for conversion.

In the case of serial code conversion using a circuit as shown in Figure 3.88, the timing sequence can be ended after the transmission of the LSB. For this reason, however, it must be as long as the longest code word to be converted. Today, the simplest method of converting Gray code back into BCD code is to use a programmable read-only memory (PROM).

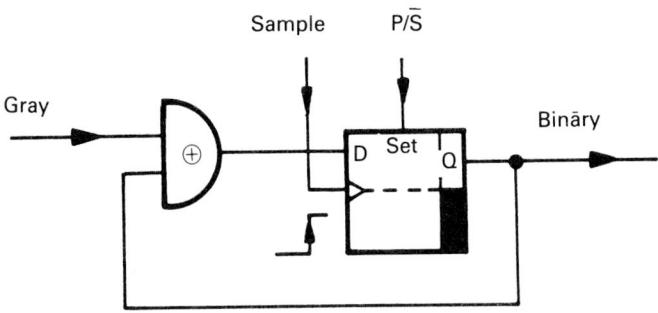

Figure 3.88 Serial code conversion, Gray to binary.

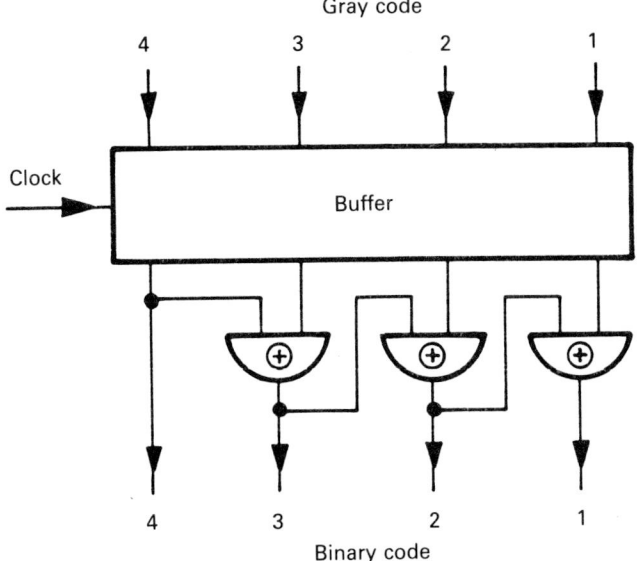

Figure 3.89 Parallel code conversion, Gray to binary.

Table 3.7    Formation of BCD Gray codes

| Decimal no. | Binary code | Gray code | Reflexion 1. | 2. | 3. | 4. |
|---|---|---|---|---|---|---|
| 0 | 00000 | 0 | | | | |
| 1 | 00001 | 1 | | | | |
| 2 | 00010 | 1\|1 | | | | |
| 3 | 00011 | 1\|0 | | | | |
| 4 | 00100 | 1\|1 0 | | | | |
| 5 | 00101 | 1\|1 1 | | | | |
| 6 | 00110 | 1\|0 1 | | | | |
| 7 | 00111 | 1\|0 0 | | | | |
| 8 | 01000 | 1\|1 0 0 | | | | |
| 9 | 01001 | 1\|1 0 1 | | | | |
| 10 | 01010 | 1\|1 1 1 | | | | |
| 11 | 01011 | 1\|1 1 0 | | | | |
| 12 | 01100 | 1\|0 1 0 | | | | |
| 13 | 01101 | 1\|0 1 1 | | | | |
| 14 | 01110 | 1\|0 0 1 | | | | |
| 15 | 01111 | 1\|0 0 0 | | | | |
| 16 | 10001 | 1\|1 0 0 0 | | | | |
| 17 | 10010 | 1\|1 0 0 1 | | | | |
| 18 | 10011 | 1\|1 0 1 1 | | | | |
| 19 | 10110 | 1\|1 0 1 0 | | | | |
| 20 | 10100 | 1\|1 1 1 0 | | | | |
| 21 | 10101 | 1\|1 1 1 1 | | | | |

One method of using Gray code for the binary sub-coding of decimal places to several decades is the reflection process. A decade with the normal sequence of figures 0 ... 9 is followed by a reflected decade with the sequence 9 ... 0. This reflection is repeated every time a figure is reflected in the higher decimal place for the lower place. Table 3.7 shows how this BCD Gray code is created.

As the zero word (0000) is avoided, the BCD Gray excess-three code is being used more frequently (Figure 3.90).

A similar unit-distance code in which a change occurs in a single binary position even where there are several decimal changes is the Petherick code, Figure 3.91. Here a continuous progression is achieved by the fact that the subsequent decade is a contributory factor for recognition. Within each decimal place the code pattern is repeated only after 20 increments (in relation to the decade in question). The value of the fourth track of each decade is inverted if the next highest decade has an odd value, which corresponds to the complement of 9.

By controlling the highest decade (inverting track 4) in angle

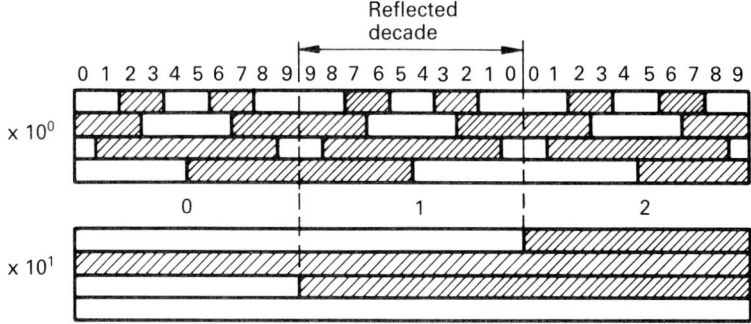

Figure 3.90   BCD Gray excess-three code.

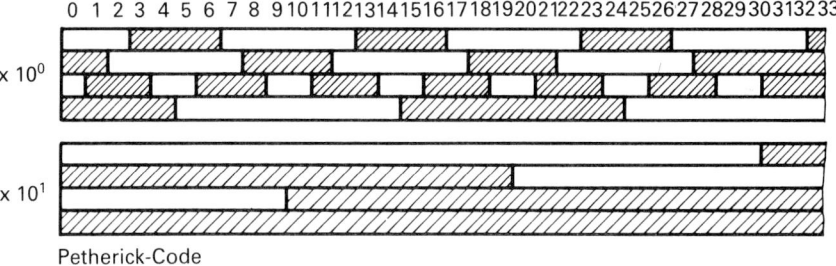

Petherick-Code

Figure 3.91   Petherick code.

encoders with the same direction of rotation forward or backward, counting can be selected [65]. The Petherick code therefore avoids ambiguities when passing from 99 to 100, 999 to 0, etc.

If, for example, an angle encoder is to be constructed for a measurement range of 3600 increments, corresponding to a resolution of 0.1°, then several positions will change their significance when passing from 3599 to 0. But three positions change simultaneously when passing from 35 to 0 (resolution 10°). In order to obtain a different range of measurement from that produced solely from second powers, an equivalent number of increments must be omitted. If the increments of the highest and lowest value are omitted the continuous progression is retained. An encoder with 360 increments is obtained by omitting $2 \times 76 = 152$ positions of the possible 512 increments of a 9-bit code.

The Gray number $(G_n)_{360}$ can be determined from the maximum Gray number $(G_m)_{512}$:

$$(G_n)_{360} = (G_{n+76})_{512}$$

The code produced is a Gray excess-76 code, where excess-76 means that the Gray numbers used have been displaced by 76 positions in relation to the full Gray code.

For 36 positions a Gray excess-14 code would be used, for 720 positions a Gray excess-152 code and for 1000 positions a Gray excess-12 code [66].

When deciding between these three methods of avoiding defective readouts with coded measurement bases for measuring angles or distances, not only must the lowest cost be borne in mind in terms of the number of code tracks, scanning elements or components, but the cost of creating a scale, which depends to a great extent on the manufacturing tolerances required, must also be considered. This last point is often the key factor in deciding on double scanning, as the permissible tolerances increase as the tracks become coarser. In contrast, in the case of unit-distance codes, all the tracks must have as narrow a tolerance as the finest track. Moreover, double scanning makes possible measuring solutions which in principle are not possible using single scanning. An example of this would be the coupling of several encoder discs by means of a step-down gear system.

### 3.8.2  Double scanning

The double scanning method is preferred for use with binary code and BCD code in order to render ineffective index errors in the measurement base and scanner assembly.

As a result of the regularity still to be considered for the distance of the scanning elements from the readout line, a V-shaped arrangement is produced in binary code, from which this process derived its name, V-scanning [67].

In this method, two scanners are provided in each code track, except for the finest one, which serves as a reference track. Both the redundancy in the code itself (synentropy) and the redundancy obtained through the position of the scanning elements are used for error correction [67;68]. The method is shown in practice using the example of a binary scale consisting of two code tracks, Figure 3.92. The readout line is specified by the contact element $A_0$ with the finest track. Opposite this are the contact elements $A_{1N}$ and $A_{1V}$ in a lagging or leading position, so that their two signals have a negative or positive phase shift in relation to the reference signal. $S_{1N}$ and $S_{1V}$ are connected with $S_0$ in the circuit and produce an output

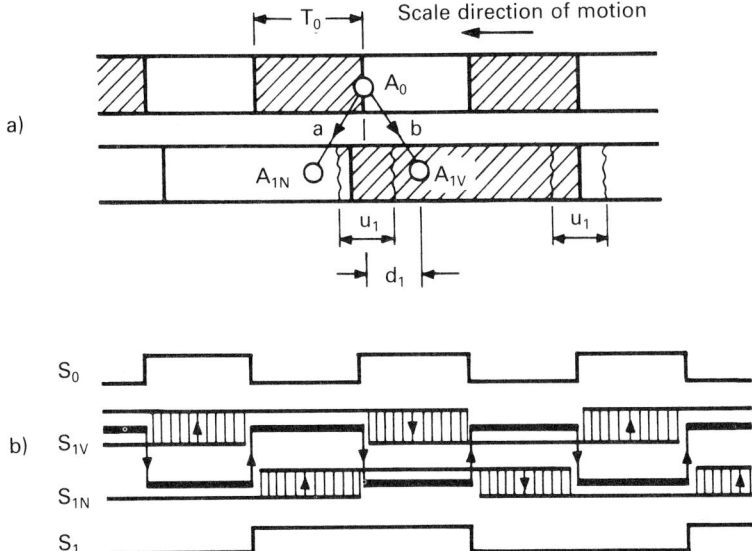

Figure 3.92 Double scanning. (a) Arrangement of scanning elements. (b) Combination of signals. $A_0$, Scanner in the finest code track; $A_{1N}$, probe, lagging position; $A_{1V}$, probe, leading position; $u_1$, uncertainty of edges of track 1; $d_1$, distance between $A_0$ and $A_{1V}$; $S_0$, signal from the 0th track; $S_{1V}$, leading signal from the first track; $S_{1N}$, lagging signal from the first track; $T_0$, division of the 0th track; a, decision when $A_0$ in shaded area; b, decision when $A_0$ in non-shaded area.

signal the edges of which match exactly in time with those of the reference signal, Figure 3.93.

The expression for this is:

$$S_1 = \bar{S}_0 S_{1V} + S_0 S_{1N} \qquad (62).$$

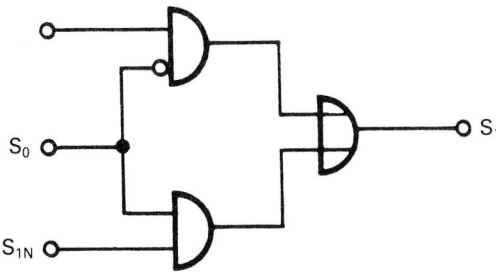

Figure 3.93 Basic circuit of V-logic.

Figure 3.92a shows, by means of the extremely extended change-over signal, which areas of $S_{1V}$ and $S_{1N}$ have been recognized. The shaded areas have no bearing on the result; they can therefore have any desired property.

Neither the position nor the shape of the edges of the subsidiary signals have any influence on $S_1$ in this area. Even oscillations from the pre-amplifiers can occur during the transition from one switching condition to another without doing any harm. The steepness of the edges of the resulting signal $S_1$ is determined solely by the characteristics of the reference signal and the logic elements used. When integrated logic elements are used, e.g. in CMOS technology, their response times as compared with the response times produced by the scanning movement play only a secondary role, so that in practice there is synchronism between the reference signal and the resulting output signal. In binary position sensing devices with more than two tracks, the process can be applied to each track if the output signal of the preceding track is used as a reference signal. In this way coded position information is obtained in which the position of the edges of all the tracks are finally determined by the finest track and a synchronous exchange of signals is guaranteed in all the tracks affected. The switching of the signals of a binary five-track code screen to synchronization serves as an example, Figure 3.94a. For the finest track, a switching amplifier with hysteresis is provided to produce edges which are as steep as possible and to avoid undesired oscillations. The mere fact that an expensive switching amplifier with a specified hysteresis is required for a single track justifies the use of double scanning in preference to the selection of a unit-distance code with single scanning, as in the latter case each track incurs the same expenditure. The considerable advantage of double scanning, however, lies first and foremost in the more accurate observation of the tolerance zones for the subsidiary signals (shaded areas in Figure 3.92b). The size and position of these tolerance zones are determined by the phase position of the two subsidiary signals $S_{1V}$ and $S_{1N}$ in relation to each other and to the reference signal $S_0$. The phase position of the subsidiary signals is, in turn, primarily governed by the geometrical arrangement of the scanning elements, i.e. by the distance of the scanners $A_{1V}$ and $A_{1N}$ from the ideal scanning line which is prescribed by $A_0$. The maximum certainty of a perfect measurement result is obtained when $d_1 = \pm(T_0/2 + n\, T_0)$ is selected (Fig. 3.92a) (where $n$ is any integer). Leaving aside any error sources, the two subsidiary signals are therefore 90° out of phase. This means that their edges are exactly at the centre of the tolerance zones and are

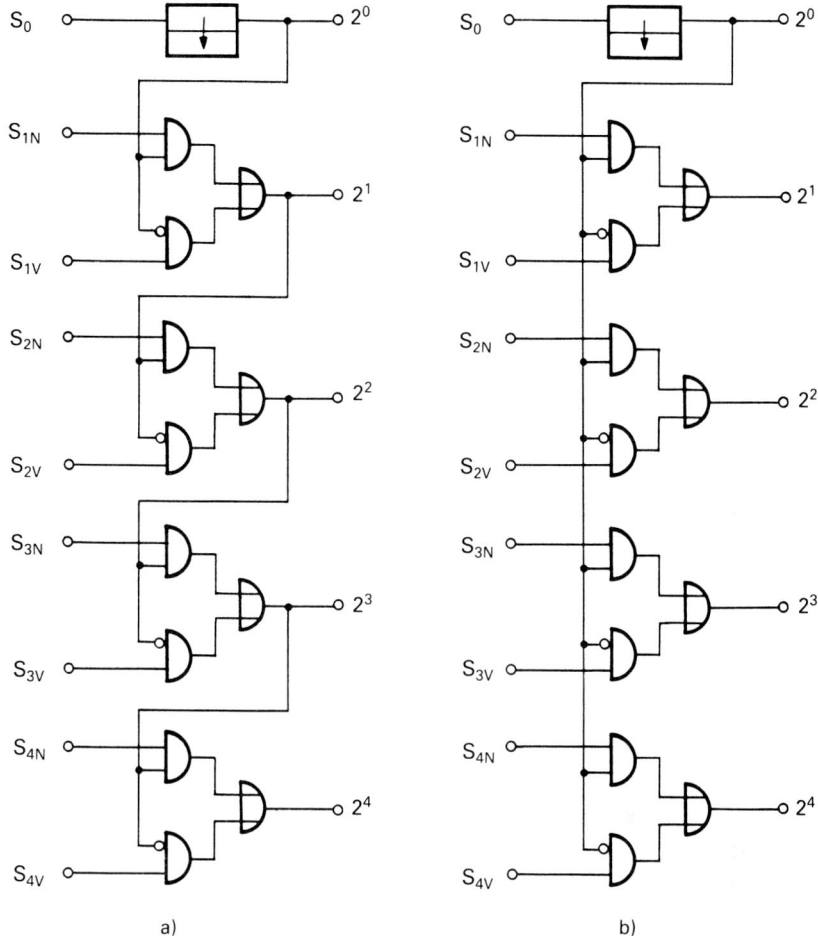

a)                                                  b)

Figure 3.94   Circuit for analysing signals from a twice-scanned binary coded scale. (a) V-logic. (b) U-logic.

enclosed by the edges of the reference signal. In Figure 3.92b, the position of the signal edges is denoted by arrows inside the shaded areas. In practice, numerous error sources, such as division inaccuracies, differences in sensitivity in detectors and amplifiers, fluctuations in line voltage and temperature and, in optical systems, fluctuations in the luminous intensity of the lamps and changes in the distance between the scale and the scanning element, are unavoidable. These errors do not influence the scanning result provided that they do not produce a phase shift in the subsidiary signals greater than $\pm T_0/2$. Figure 3.95 shows vividly the effect V-logic has on the selection of

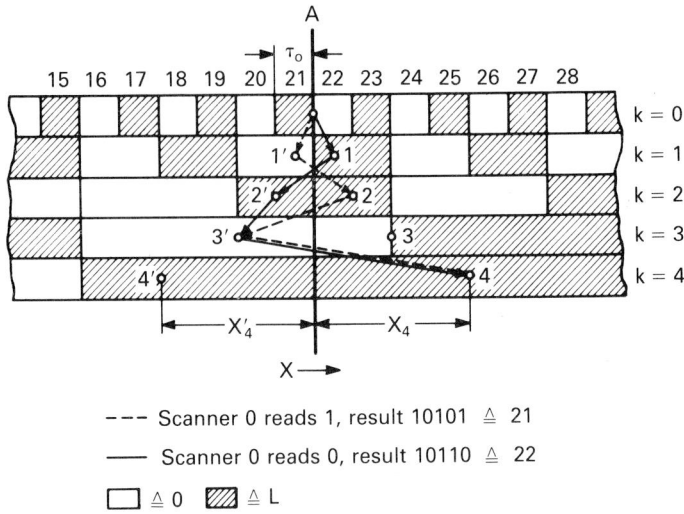

Figure 3.95   Using the double scanning principle on a binary coded scale. *A*, Scanning line; $t_0$, division of the finest track; x, direction of motion; *k*, order of code tracks.

scanning points. As the limit of resolution of a measurement process is determined by the smallest summing error, the cost of the technology necessary to achieve the finest tracks possible in a digital measurement process must be accepted, while at the same time working with the greatest possible degree of certainty, i.e. with the subsidiary signals 90° out of phase. If, in the subsequent coarser tracks, the requirements for the components used and the manufacturing tolerances are gradually reduced, it becomes apparent that although some of the errors increase at first, they then taper off towards a final limiting value. This means that, for example, a larger error occurs in track $2^3$ than in track $2^0$ and that this error is perhaps two or three times larger than the error in track $2^0$, but nothing like eight times. An even smaller fraction of the possible tolerance zone is used as the significance of the tracks increases. The phase shift between the subsidiary signals can then, if required, be reduced. In measurement processes with a large range of measurement, it is useful and often necessary to reduce the distance between the scanning elements relative to the division of an individual track in order to prevent the scanning devices taking up too much room.

In practice, a large enough phase shift is selected to keep a margin of certainty for the maximum region of uncertainty.

## 3.8.3 Special forms of double scanning

In addition to the V-shaped arrangement of scanning elements described, Figure 3.96a, a U-shaped arrangement is also possible in binary code, Figure 3.96b. In this case, the selection logic is simplified (Figure 3.94b) as, depending on the uncorrected result of the finest track, either all the leading or all the lagging scanners are analysed. Against this advantage is the disadvantage that only half of a division of the reference track is available as the maximum tolerance for all the signals. This method is therefore only used where narrow tolerances do not cause increased technical costs. On the other hand, there are cases where double scanning is only possible with U-logic. An example of this is Gray code, Figure 3.96d. Here, the scanning elements are reversed according to the uncorrected result from an auxiliary sensor $H$ which is placed half a division away from the main sensor in the 0 track. The scanners for the remaining tracks are now placed so that they lead or lag the main sensor which prescribes the read-out line by a quarter of a division of the finest track. As the division of the finest track in Gray code corresponds to two counting increments, the absolute amount for the permissible displacement of scanners is the same as in binary code with U-scanning.

The rule for scanning is that when auxiliary sensors are placed in a lagging position and logic 0 is read at the auxiliary sensor, the lagging signal is analysed in the subsequent track and the leading signals in the remaining tracks. Correspondingly, the reverse is true when logic 1 is read at the auxiliary sensor. As Gray code is a unit-distance code, double scanning is only of relatively limited benefit. An exception to this is when several discs are coupled. Things are however different with BCD code, Figure 3.96c. If first of all the four tracks in a decade are observed, e.g. the 0 decade, it will be seen that the signal width is only doubled from track to track up to track $2^2$. Track $2^3$ is only as wide as two units of the 0 track. For this reason tracks $2^1$ and $2^3$ must be controlled by track $2^0$. On the other hand, track $2^2$ is controlled by track $2^1$. In BCD code therefore, a combination of V- and U-scanning is required within a decade. If the first three tracks are observed there is V-scanning only, but when, in contrast, tracks 0, 1 and 3 are viewed in isolation we find U-scanning.

A second peculiarity of BCD code is the transition from one decade to the next. Again, the narrow width of the last signal in each decade causes problems. As the maximum tolerance for the displacement of the scanning elements depends on the width of the preceding

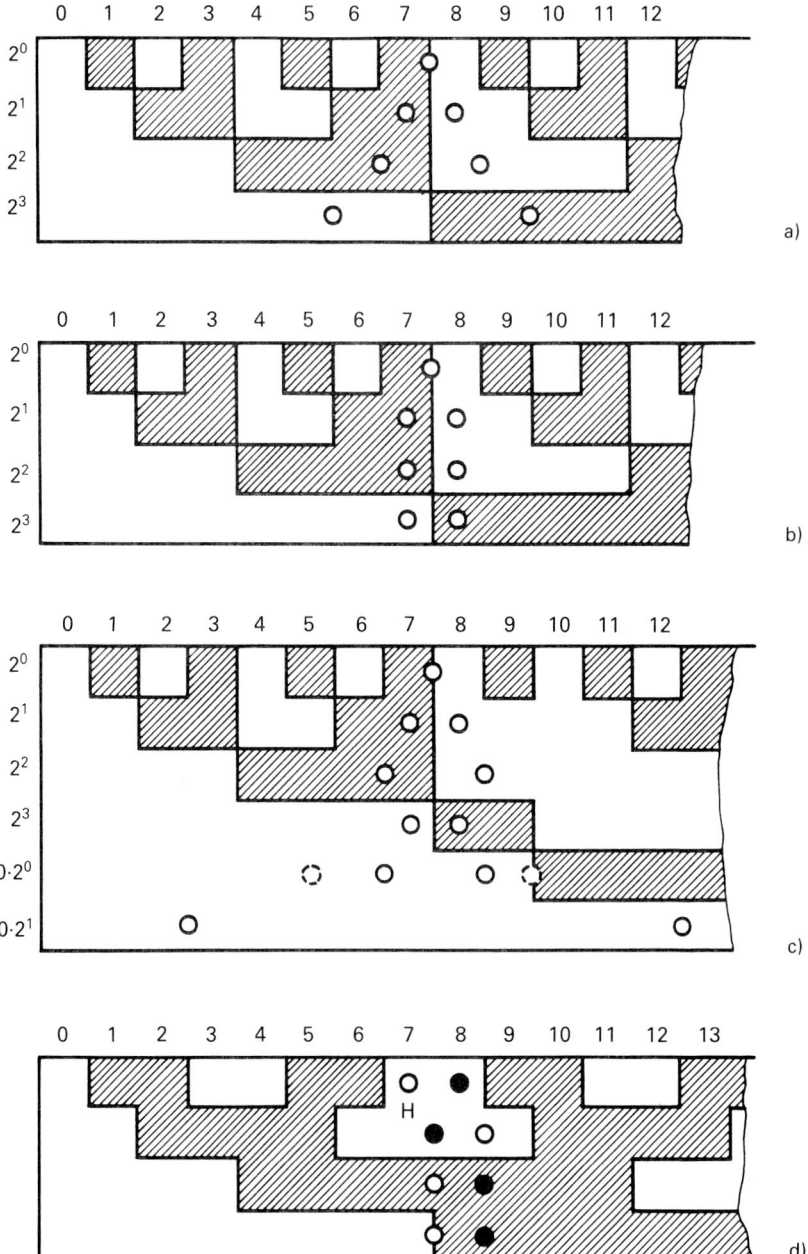

Figure 3.96 Arrangement of scanning elements for double scanning. (a) Binary code with V-scanning. (b) Binary code with U-scanning. (c) BCD code with modified V-scanning. (d) Gray code with U-scanning and auxiliary probe H.

track, and this amounts to only $1 \times 2^0$ from $1 \times 2^3$ to $10 \times 2^0$, an auxiliary signal is usually produced to control the first bit in each decade [69]. The greatest range of tolerance is obtained if the signal $5 \times 10^{m-1}$ is decoded, thereby controlling track $1 \times 10^m$ (where $m$ is the next decade). It is considerably easier, and only involves a slight loss of tolerance, to produce an auxiliary signal through the conjunction of $2^2$ and $2^3$. In this way a scanning ratio of $4:6$ and a doubling of the permissible tolerance are obtained [70]. In Figure 3.96 two pairs of scanning elements have been drawn in track $2^0 \times 10$. The dotted one represents the use of an auxiliary signal generated from $2^2$ and $2^3$. Figure 3.97 shows the appropriate logic circuit.

The V-shaped arrangement of scanning elements in binary code can be replaced by dividing each code track into two halves and displacing the scanning elements by an amount corresponding to the original spacing, Figure 3.98. The scanning elements are then arranged in line. The code produced shows a certain similarity to Gray code. Compared with V-scanning, however, this method has the disadvantage that increased costs for producing the code scale or the code disc are not matched by savings in signal analysis.

## 3.8.4 Scanning with preselection

In double scanning two signals per code track are produced, of which only one continues to be used subsequently. This means there is not only redundancy with regard to information but also with regard to circuit technology. On the assumption that it is completely unnecessary to produce two signals if it can be decided beforehand which of the two will be required subsequently, we come to a conclusion which means expenditure, particularly on sensitive pre-amplifiers, is drastically reduced [71;72]. This is very important in high-resolution measurement processes where pre-amplifiers not only have to be highly sensitive but accurate and stable as well. In contrast, a certain amount of extra expenditure on logic elements, which nowadays are not only inexpensive but can be supplied with a high degree of integration, is of less significance.

Figure 3.99 shows a scanning arrangement in which the decision on the analysis of a signal is linked to its generation. A feature of this circuit is the allocation of two controllable light sources to each code track, with the exception of the finest track which serves as a reference track, in which there is only one light source. The second important feature is the serial scanning of the tracks which enables a

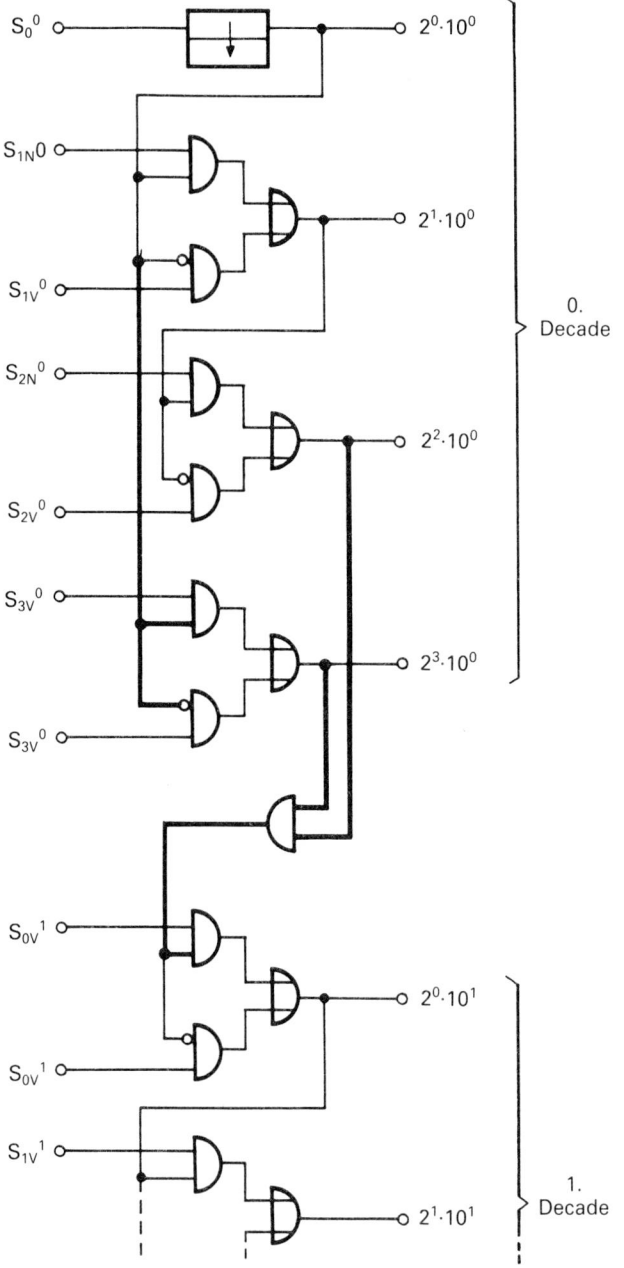

Figure 3.97   Circuit for analysing signals from a twice-scanned BCD coded scale.

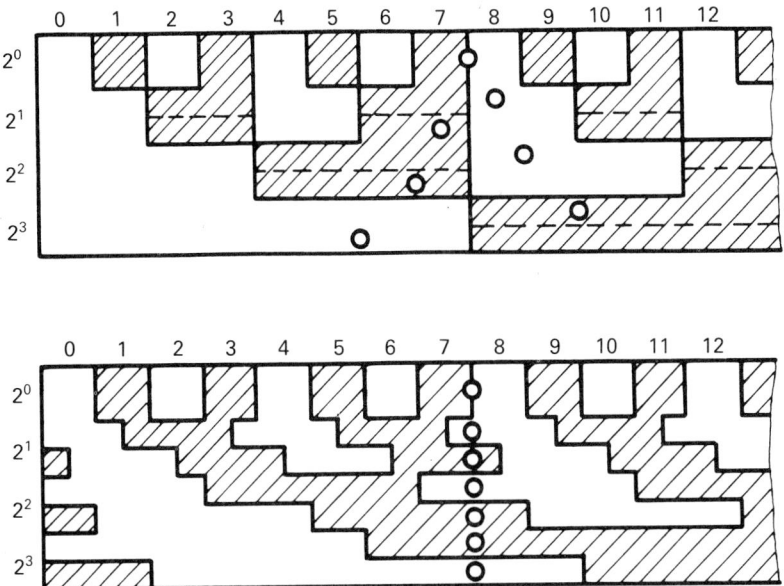

Figure 3.98   Replacing V-scanning with V-code.

single pre-amplifier to be used and at the same time permits the serial transmission of information without further conversion. The operation of the circuit can be studied most simply with the aid of the timing diagram, Figure 3.100. In the example, a four-track code disc with the information 1011 (decimal 11) just appearing is to be scanned. Synchronization of the measurement process and the data processing unit is achieved by means of a call pulse which initiates the scanning by positioning in the D flip-flop FF1, Figure 3.99. With the next positive edge of the clock pulse signal $T_1$ a logic is entered in the shift register and at the same time the FF1 is reset via the output A of the shift register SR.

Controlled by the clock pulse signal $T_1$ this 1 'roams' through the shift register so that the four tracks on the code disc are scanned via the signals of outlets B–D one after the other. The decision as to which of the two light sources (light-emitting diodes) in each track will be activated depends on the position of the FF2, on to whose D output the signal preceding it in time, which matches the track with the next lowest significance, is switched. To suppress interference pulses, which are caused by gate transit times, rise and decay time constants of light sources, photodetectors and pre-amplifiers, output storage is provided in the form of the FF3, which is supplied with

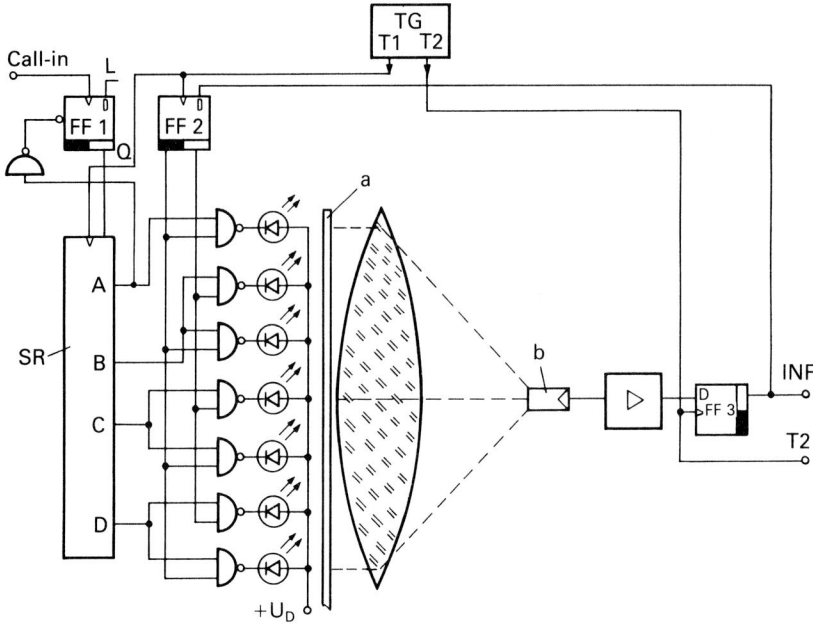

Figure 3.99 Scanning arrangement using controlled light sources. a, Code disc; b, photodetector; SR, shift register; $A$, $B$, $C$, $D$, shift register outputs; FF1, D flip-flop 1; FF2, D flip-flop 2; FF3 D flip-flop 3; $U_D$, diode voltage; TG, timing generator; $T1$, clock pulse 1; $T2$, clock pulse 2; INF, information.

clock pulses of signal $T_2$. Since $T_2$ lags $T_1$ by a quarter period, a specific time is allowed to set the appropriate value. As a result of the control of the selection circuit by the preceding bit with the next lowest significance, the bit with the lowest significance is transmitted first in serial transmission and the bit with the highest significance is transmitted at the end.

## 3.8.5 *Absolute shaft encoders*

An absolute shaft encoder incorporates a rotatable coding disc and a readout device. The principle of an optical angle encoder is shown in Figure 3.101. The code disc is a glass plate with code patterns photographically engraved. Transparent regions give logical 1 signals while blanked off regions give logical 0 signals. A lamp with a lens to produce parallel light rays illuminates the disc at one point, opposite the

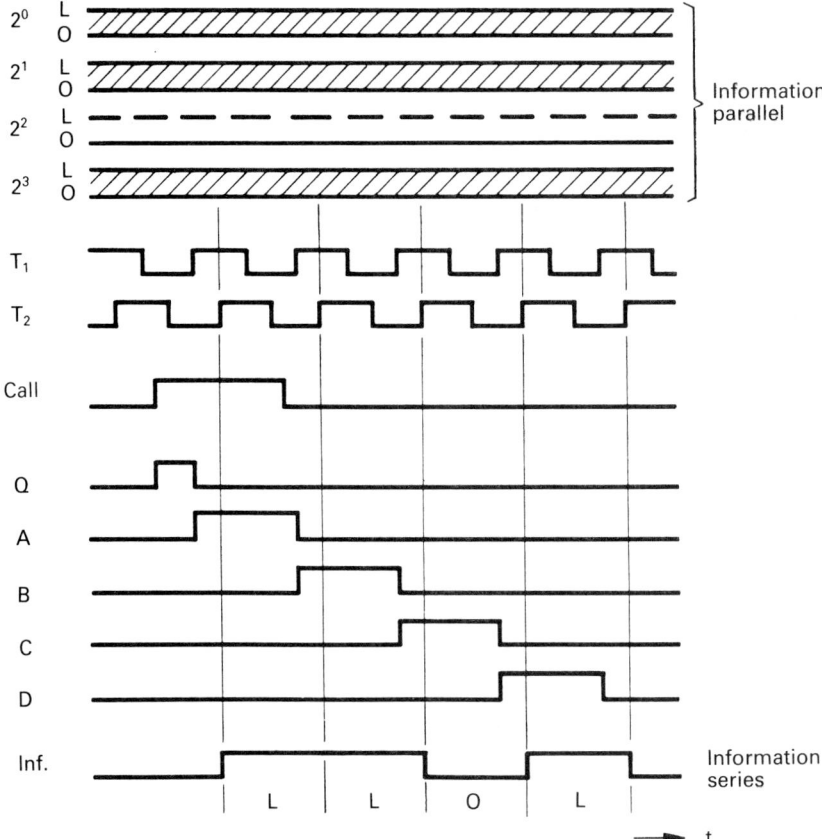

Figure 3.100   Signal diagram for interrogating coded scale. $A$, $B$, $C$, $D$, Shift register signals.

position of the readout device with its photoelectric cell. Coded information results from preamplification.

Examples of a code disc and the readout device are shown in Figures 3.102 and 3.103.

To keep the number of code tracks to a minimum, the disc is not in actual BCD code, but in a special code of its own. In this, several tracks are read up to six times, and the BCD code is obtained through exclusive combined decoding. The code tracks on the disc are determined not only from their value, but also from the most useful possible layout of the readout element, which must be arranged in a relatively narrow space. The first and finest track of the code disc consists of 1000 light and dark fields, of respective relative

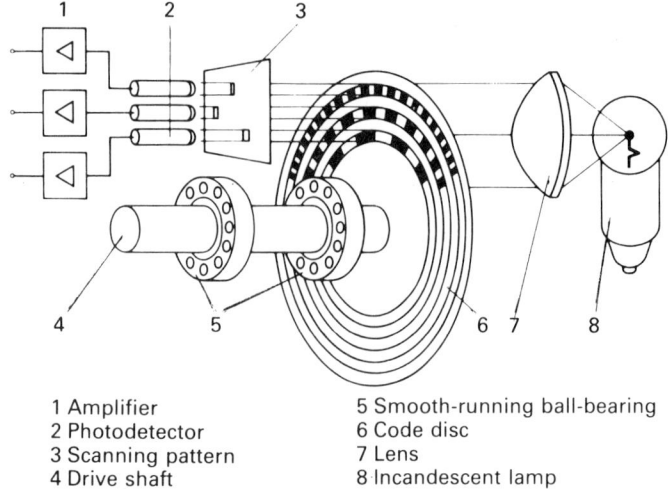

1 Amplifier
2 Photodetector
3 Scanning pattern
4 Drive shaft

5 Smooth-running ball-bearing
6 Code disc
7 Lens
8 Incandescent lamp

Figure 3.101  Basic representation of a photoelectric angle encoder.

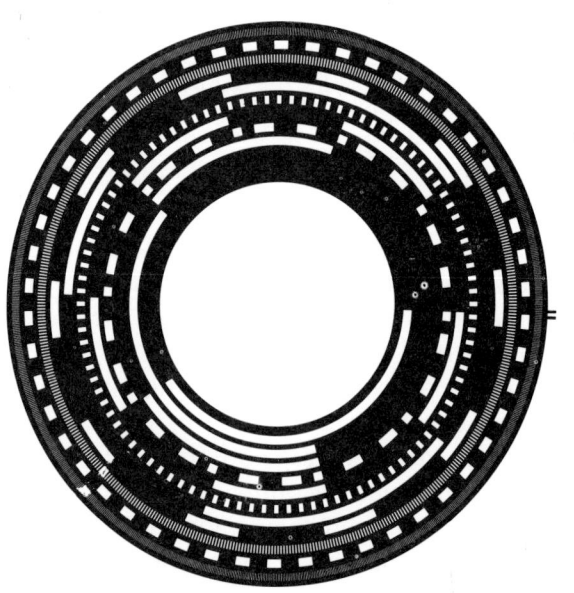

Figure 3.102  Code disc. (Picture courtesy of AEG-Telefunken.)

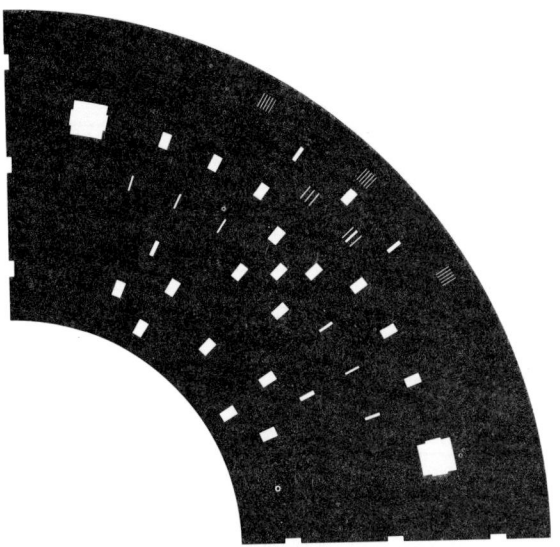

Figure 3.103   Scanning pattern. (Picture courtesy of AEG-Telefunken.)

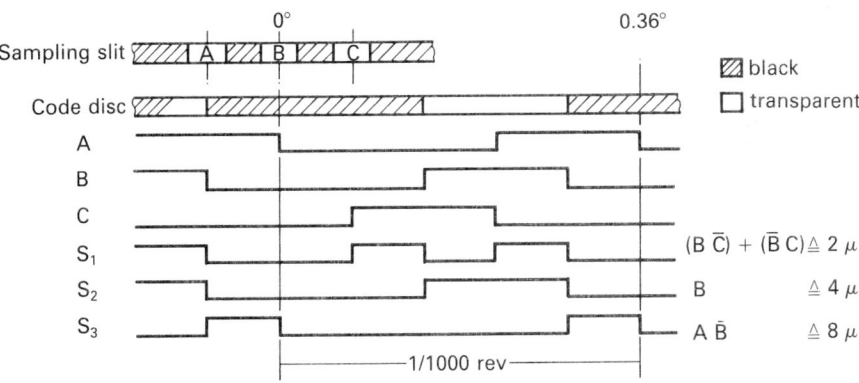

Figure 3.104   Triple scanning. *A*, *B*, *C*, Photodetector signals; $S_1$, $S_2$, $S_3$, combined signals.

widths 4:6. Figure 3.104 shows a section of this track with a readout made up by the three readout signals *A*, *B* and *C*. The signals corresponding to this readout position are therefore set to $\frac{1}{5}$th of the signal period. Their combination produces three signals in BCD code, of which the finest $S_1 = 1/5000$th of a revolution, which by

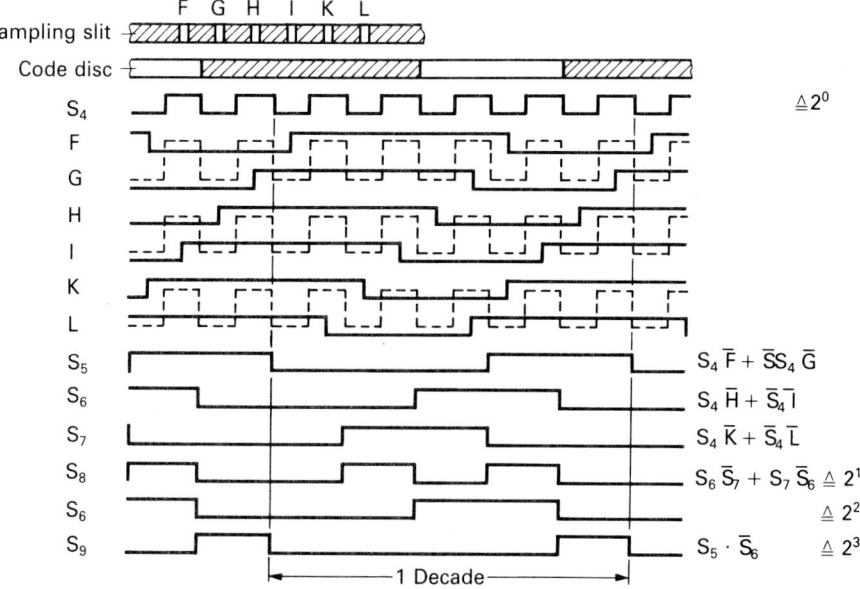

Figure 3.105   Combination of V-scanning and multiple analysis (for 3 bit). *F–L*, Photodetector signals; $S_5$–$S_8$, combined signals; $S_4$, reference signal.

replacing in a translatory movement over a measurement spindle with 10 mm slope, corresponds to $2\,\mu$m. $S_2$ then corresponds to $4\,\mu$m and $S_3$ to $8\,\mu$m. The next coarsest track (no. 3 on the disc) is read twice, which in relation to the evaluation of signal $A$ produces only one signal (V-logic).

Figure 3.105 shows a track read six times. Here, two principles are combined. One is that from this track three bits of a BCD coded decade are obtained, which requires three readouts, and the second is that the signal flank lies exactly in the time frame given by signal $S_4$. This is permissible only through the use of one pre- and one post-auxiliary signal for each main signal, with corresponding evaluation using V-logic, where the number of readout positions is doubled from six. If several rotations are absolutely measured with an angle encoder, then further code discs are coupled with reduction gearing; two such possibilities are shown in Figure 3.106.

In this case, the gearing is so arranged that the input shaft which carries code disc 1 is multiplied by a factor of 20 for code disc 2 and a factor of 400 for code disc 3. The problem of simultaneous signal exchange in all tracks of comparable discs which arises from this can be solved by double readout and V-logic. Indeed, it can be said that

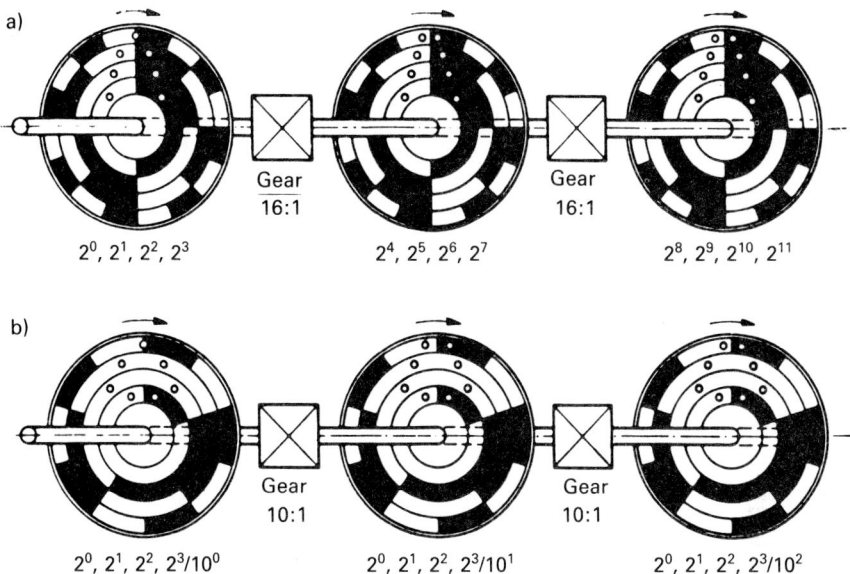

Figure 3.106 Basic structure of a binary coded and a binary decimal coded angle encoder with a measuring range of more than one rotation. (a) Sixteen-part code disc. (b) Ten-part code disc.

only in the context of different measurement systems, as with three code discs, does V-logic really show its full power. This is clearly shown in the example of the transfer from the first to the second disc. Because of the double reading, the controlling tolerance is that the controlled track must be set as a maximum to about half of the smallest signal from the previous track [73]. In transferring from disc I to disc II, the control signal uses a combination of 8 and 4. The tolerance range is therefore in the region of ±72°, see Figure 3.107. For the tolerance of the gearing, a ratio of $1:20$ is a solution for the

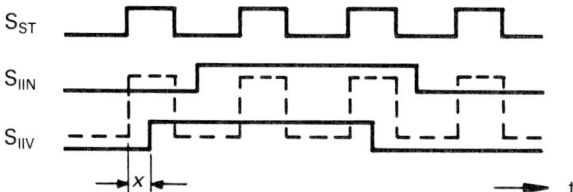

Figure 3.107 Changing over from code disc I to code disc II. $S_{ST}$, Control signal from code disc I; $S_{IIV}$ and $S_{IIN}$, leading and lagging signal; $x$, range of tolerance.

input shaft with ±15° for the fixed output shaft, so that a normally-toothed spur gear without preloaded gear pairs can be applied. Code disc II is adjusted by the mounting of code disc I. Here there is a tolerance of ±0.5° or ±10°, relative to the input shaft. The readout uncertainty for the second disc resulting from the finite width (0.1 mm) of the readout slit leads to an error of ±15°, also arising from the input shaft.

The total tolerance is therefore ±40°. For safety, it is the difference between this and 72° and so it is ±32°.

Up to now it has been assumed that V-logic permits an exactly simultaneous exchange of signals in all tracks. However, switching V-logic itself requires a finite time, which is the sum of the propagation delay times of the gates times. In order to synchronize the interrogation of measurement values through a data processing system, alteration of the measured values during transfer through an interrogation signal must be prevented. A switching of V-logic permits such prevention with safety, when a fixed number of signals beginning with the lowest value is taken from the buffer store by V-logic, Figure 3.108. The number of inhibited signals depends on the maximum possible rotary speed, the resolution and the interrogation period. The buffer store is inhibited by the interrogation signal for the duration of the interrogation period so as to take in new information.

The suppression of the effect of gear play between two or more coupled angle encoding discs can also be achieved within certain limits without the use of V-logic, if a microcomputer is integrated with the angle encoder [74]. This microcomputer simulates the V-readout, more accurately described as a U-readout, in which, in relation to the measured result of the foregoing disc and an auxiliary track of the disc in use, the readoff value is clarified as valid or corrected by 1. The use of an auxiliary track has already been dealt with in section 3.8.1 in Figure 3.86. Figure 3.110 shows two gear-coupled coding discs. On disc 1 in the example 100 positions per revolution are available, so that the clarity of half the code is expressed with decimal numbers. It is interesting here to note especially the transfer from 99 to 0, the so-called zero transfer. Disc 2 consists of an auxiliary track, corresponding to one revolution of disc 1. Disc 2 is arranged with a ratio of 16 : 1 to disc 1. Its other tracks are in Gray code, so that single readout applies and no erroneous intermediate values in the transfer between one code and the next can arise.

For the adjustment of the discs relative to one another in the example, the second disc is mounted at a lag angle $\delta$. During the

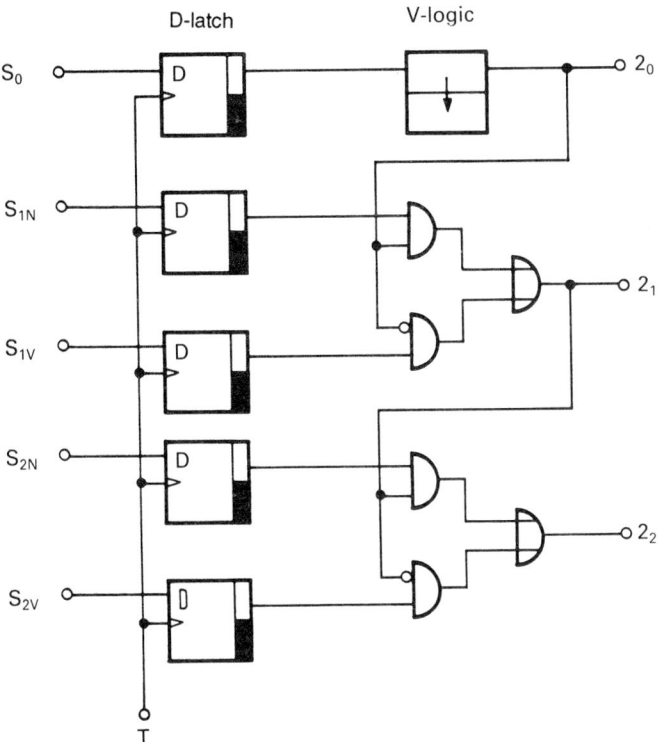

Figure 3.108   Syncronizing the position values of the absolute shaft encoder at a fixed point in time.

interrogation of disc 2 the auxiliary track differs from the actual value of disc 1 over the evaluation. The bit with the lowest evaluation of disc 2 must now change its value, if the auxiliary track is in a defined position (in the example on a transparent field) and the first disc has made a full revolution. The conclusion of one revolution in the example is represented by a transfer from 0 to 99 or the reverse. The output now indicates that with a black field in the auxiliary track the read off value applies. If a white field shows, the position differs additionally from disc 1. If this is before the zero transfer, then it applies for the value read off from reading line A. After the zero transfer, the value conforms to a fictitious reading line A′, which is obtained as a result of the readout adding 1 to the reading from A. The process also works with a lead adjustment of disc 2. In this case the 0 transfer of disc 1 must be subtracted from that of disc 2.

A property fixed by the principle of absolute measurement is the zero point established as invariable on the scale.

A change in the position of the zero point is only possible with an absolute shaft encoder through two processes: the mechanical rotation of the code disc on the measurement spindle or other rotary machine part to be measured, or the subtraction or addition of a correction factor to the measured value in a computer. Both processes can be carried out today with angle encoders: mechanical adjustment can take place via a friction clutch, and electrical corrections can be done using built-in microcomputers, Figure 3.109.

In connection with a non-volatile buffer (EEPROM), the actual value for the code disc can be stored for each arbitrary time point, and subtracted from the measured value for future measurements. This allows the zero point of the absolute shaft encoder to be set in any preferred position via a push button, which up to now could only be carried out using an incremental shaft encoder.

A further application of the microcomputer is self-checking, by which the operational serviceability can be checked. In this is a stochastic check, which checks measured values to see if they lie within the measurement range. The writability of sender data can also be checked (monotony checking). In relation to the known output code, the microprocessor takes over the code translation, as well as the use of test bits.

The direction of rotation and the readout of the shaft encoder are also programmable. One difficulty that is especially encountered with high resolution angle encoders is data transfer, which must carry many parallel signals and thus requires multicore cables. Multicore cables are less flexible than cables with fewer cores, and so are more prone to mechanical effects. As a solution, serial data transfer is

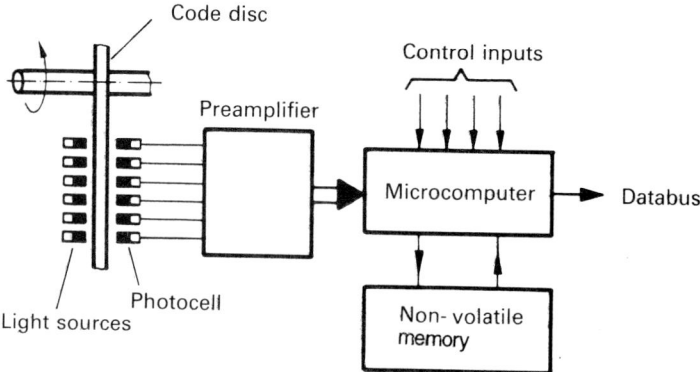

Figure 3.109   Angle encoder with microcomputer.

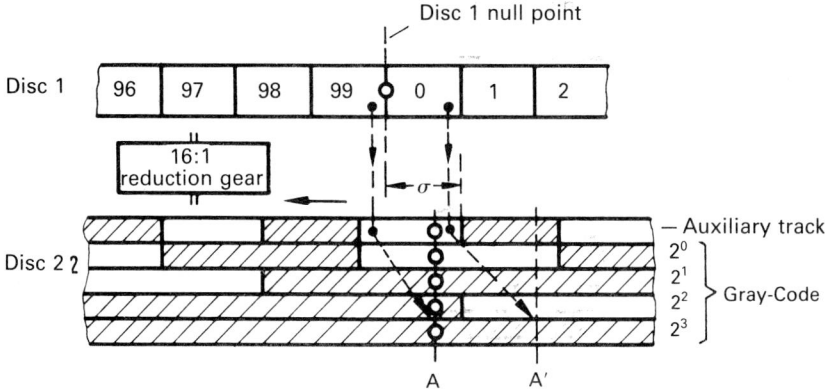

Figure 3.110   Scanning mechanically coupled angle encoder discs using an auxiliary track.

possible. Control of serial transfer is a typical task for a microcomputer. The application of such an intelligent measurement system to various outputs is possible with the software of microcomputers, and the possibilities will certainly be improved and extended in the near future.

### 3.8.6   Linear encoders

A linear encoder is a direct and digitally absolute position measuring system for length measurements [75]. Its main application is found where accuracy and security of measurements are essential, e.g. trial bores and measurement machinery.

It is limited through its high production costs, and the high operational safety and reliability of incremental linear measurement systems now means there is no difference between them.

In its material dimensions and interrogation, the same physical laws apply as for angle encoders. In practice only optical readout systems are applied, with transmitted-light glass scales first coming onto the market, followed by gold-plated steel scales operating via reflected light. Glass scales could be produced in lengths up to approximately $2\,m$ with a resolution of $5\,\mu m$. Metal scales have a resolution down to $2\,\mu m$ can be made in lengths of several metres. Coding can be either in binary or BCD. High resolution scales must be made in one piece for a given measurement length, since it is not possible to obtain combined numbers from partial scales due to unavoidable production tolerances. If a linear encoder is compared with the commonly used angle encoders, then the disadvantages of the linear encoder are seen to be the much more invalid dimensioning of the measurement base.

### 3.8.7 *Mechanical digital position sensors*

In many processing machines, adjusting spindles for stops, supports, etc. are set by hand. However, an accurate setting of the desired position can only be obtained using mechanical digital position sensors, Figure 3.111. These consist mainly of a rolling toothed wheel and a

Figure 3.111   MPA 90 mechanical position indication. (Picture courtesy of Erwin Halstrup Multur GmbH.)

measurement system. The rotation of the spindle is transferred to the rolling toothed wheel via a hollow shaft. In contrast to analogue displays with scales and pointers, digital indicators are characterized by greater accuracy and ease of readout [76]. Depending on the ratio chosen, which can be either metric or imperial, and any desired magnification, rotary speeds between 150 and 300 r.p.m. are possible. These display systems are therefore equally suitable for machine controlled spindles. Maximum permissible rotary speed depends on application. The highest resolution is 0.01 mm for a spindle magnification of 1 mm. Since the device is fixed to the shaft by a hollow shaft, and the casing must only be prevented from rotating by a pin, angular errors or unevennesses of the machine measurement surface do not cause problems. End shields can be applied with these devices without interfering with their function.

## 3.9  Distance measurement by the pulse-time delay method

The measurement of displacement between two points can be carried out by timing a pulse. The equipment needed consists of a pulse sender and a pulse receiver, together with time measurement equipment. In the simplest case, the pulse sender is located at one end of the distance to be measured and the receiver at the other. In this process the energy applied varies with the square of the distance. In many cases, however, it is necessary to have sender, receiver and time measurement equipment in one place and the displacement measured via reflection of the pulses from the object in question. Sometimes, the use of special reflectors is possible. In most cases, however, the surface of the object in question must be used as a reflector. In reflection processes, the energy employed falls with the fourth power of distance. The advantage of these processes is that they allow measurement of both moving and what are referred to as 'uncooperative' objects, such as, for example, the distance between a radar installation and an enemy aircraft.

Radio distance finding systems can also include those in which a moving object is equipped with an answering transmitter which responds to a primary frequency with a secondary frequency very slightly displaced from it. The measurement range can therefore be greatly increased for a given transmission energy, because in this case energy falls off with only the square of the distance. Depending on application, different frequency spectra may be used: ultrasound, electromagnetic radiation from medium to microwave wavelengths or light.

In every case the law distance = time × velocity applies. The best known examples of this principle are the echo sounder, radar, radio distance measurement, laser distance measurement and, most recently, ultrasonic distance measurements in machine tools. Important criteria for use of this technology are carrier frequency, pulse frequency, velocity of propagation, range, resolution and reliability. Basically, pulse measuring systems are accurate when the time between two measurements is greater than the total time for a pulse to travel from the sender to the reflecting object and back.

The limit of pulse frequency, which is simultaneously expressed as the parameter for the time between individual measurements, does not play a part so long as the distance to be measured between two successive measurements does not change.

The maximum range $d_{max}$, which can be simply expressed, is:

$$d_{max} \leqslant \frac{c}{2f_p} \tag{63}$$

where $f_p$ is the pulse frequency and $c$ is the propagation velocity.

In cases where moving objects have to be measured, pulse frequency should be as high as possible in order to obtain accurate readings. Readings are ambiguous when a pulse signal is sent out before the echo of the previous signal has been received. In such cases, no simple relationship between signal and echo can be established. There are, however, possibilities whereby this problem can be reduced, especially in radar systems.

Another property which is common to all time lag measurement systems is the existence of a lower as well as an upper measurement range. The lower limit in systems in which both sender and receiver use a common antenna, i.e. in radar systems, is limited by the pulse width of the sender and the time necessary for switching between sending and receiving. The sender can be switched only after a pulse duration $\tau$ and the receiver within a switching time $\Delta\tau$. For a given velocity of propagation, the shortest measurement range is therefore:

$$d_{min} \geqslant (\tau + \Delta\tau)\frac{v}{2} \tag{64}$$

for a reflective or two-way process [59]. For one radar pulse of $1\,\mu s$ duration, the wave front is propagated at the velocity of light ($c = 2.9979 \times 10^8\,m\,s^{-1}$), and so travels around $280\,m$ in this period of time. The shortest measurable distance in this case is therefore 140 m. If the switching time $\Delta\tau$ is also a microsecond, then the minimum measurable distance becomes 280 m.

The accuracy of the determination of the wavefront of a pulse depends on the signal-to-noise ratio, the properties of the threshold switch and the slope of the wavefront, especially on small signals which overlay the signal in use with a positive or negative noise amplitude. With laser distance measurement equipment, used for motion measurement, these factors can produce an error of between 1 and 2 m [77].

The upper limit for time lag measurements is determined by sender energy, sensitivity of the receiver, the medium in which the pulse is generated and the reflective properties of the object to be measured.

Resolution is determined by the properties of the time measurement system.

The electrical conversion of time lag $t_L$ can be either analogue or digital. If constant d.c. switching is used during the lag time, a pulse-width-modulated signal is produced, in which pulse width is proportional to range $x$. If this pulse is later integrated, then the amplitude of the output signal of the integrator is a measure of the displacement. A digital output signal allows the use of gate switching for the duration of the lag time, in order to obtain numbers which are exponentially produced by a quartz-stabilized counter. The count is then a measure of the lag time and thus of the distance.

### 3.9.1 Ultrasonic length measurement systems

A distance measuring system using ultrasonic pulses and measuring the time lag to a receiver can measure length up to 10 m [77;79;80].

It consists of a scale with a sender for current pulses and an ultrasonic receiver, Figure 3.112, as well as a separate electronic instrument for controlling the measurement head and for measurement analysis.

An important component of the scale is a tubular magnetostrictive wire, which serves as a waveguide for the ultrasonic pulses. By magnetostriction we mean reversible alteration of the geometric dimensions of a ferromagnetic body under the influence of a magnetic field. The scale consists of a stable, non-magnetic metallic tube, with the magnetostrictive waveguide with a copper wire electrical connection located at the centre. To avoid undesired reflections the end opposite the ultrasonic receiver has a damping zone of about

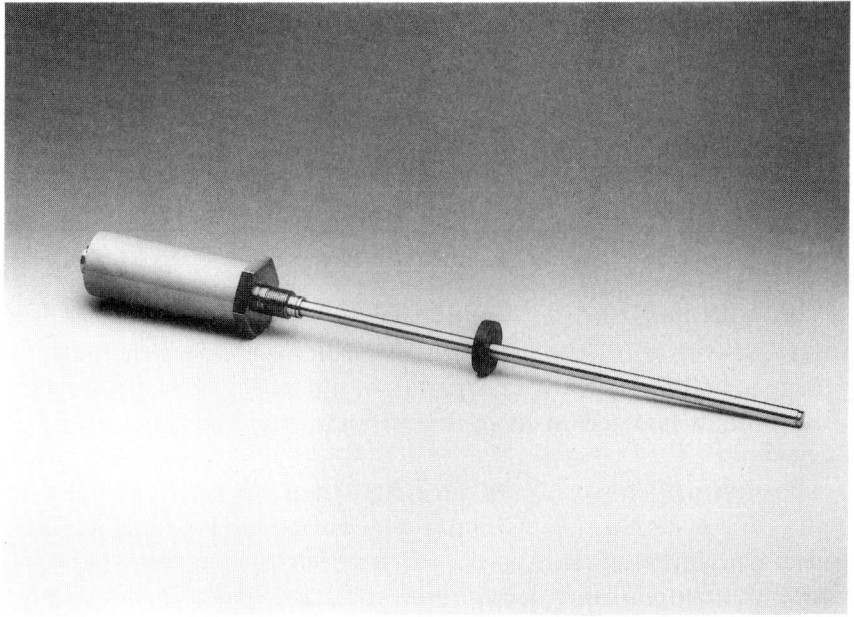

Figure 3.112   Ultrasound linear measuring system. (Picture courtesy of Balluff.)

150 mm in length. For use as an ultrasonic pulse sender and thus as a position indicator, there is an annular body in which four permanent magnets are located, Figure 3.113.

This ring magnet is fastened to a machine component, the position

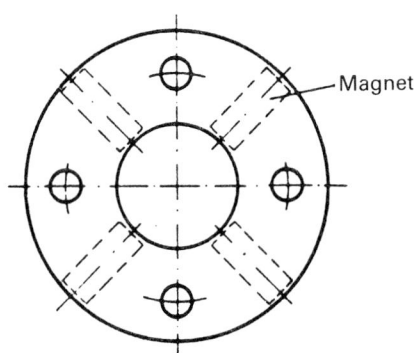

Figure 3.113   Position sensor.

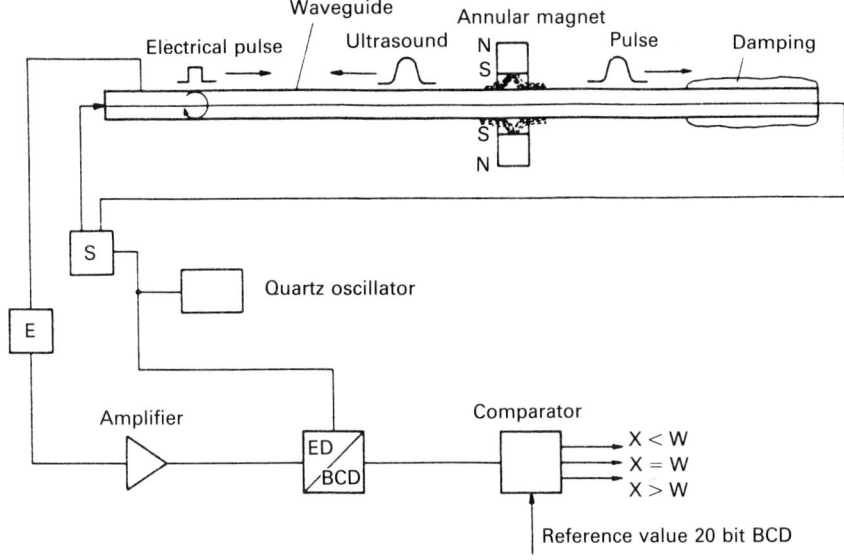

Figure 3.114   Block diagram of ultrasound linear measuring system.
(Picture courtesy of Philips.)

of which can be measured. During measurement it moves relative to
the measurement tube and its magnetic field affects the waveguide.
For measurement, a current pulse of high intensity is sent out which
propagates along the waveguide with the velocity of light, Figure
3.114. It is accompanied by an annular magnetic field. If this annular
magnetic field of the pulse coincides with the longitudinal field of the
position sender, a spiral field results. This spiral field exerts a brief
torsion effect on the magnetostrictive wire, the origin of which lies at
the centre of the magnetic position sender, Figure 3.115. The mechanical torsion pulse sets up, in both directions an ultrasonic wave in
the waveguide with the velocity of approximately $2800\,\mathrm{m^{-1}s}$ [78].
The damping equipment on the side opposite the ultrasonic receiver
cancels out the pulse in that direction. The time lag of the second
pulse between the position sender and the ultrasonic receiver is
measured. This measurement includes the lag of the electrical pulse
from the sender to the annular magnet, which is also proportional to
the distance to be measured. Since the velocity of propagation of the
electrical pulse is 10 orders of magnitude greater than that of the
ultrasonic wave, it can be neglected. The determination of the lag
itself can be carried out either digitally or analogically by the method
described in the previous section.

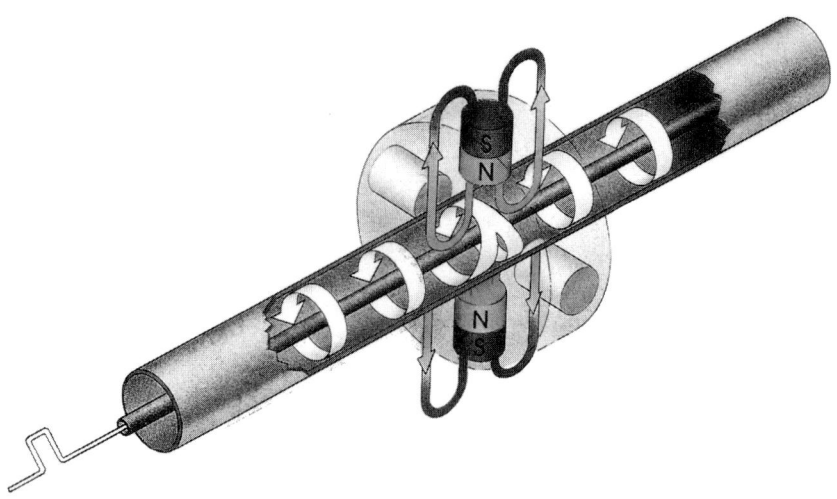

Figure 3.115   Ultrasound pulse generation. (Picture courtesy of Balluff.)

Since the sensitivity of the measurement system in general is ⩽0.1 mm, sender frequencies of ⩾28 MHz are required.

To obtain a simple measurement, a new interrogation pulse must first be sent, if the first is to be evaluated. From this is obtained a pulse frequency dependent on length. This ranges from about 9 kHz for 150 mm down to 850 Hz for 2500 mm [78;80]. It is characteristic of absolute measurement that the zero point should not alter. With ultrasonic distance measurement systems this lies with the receiver. Through the device of using two position senders, of which one is fixed, it is possible to define a variable zero point. For this, a minimum distance of 40 mm between the two position senders is necessary, which corresponds to the measured value 'zero' [79]. From these, two pulses are given, whose time lag is evaluated.

### 3.9.2   Ultrasonic measurement in air

For contactless displacement measurement systems, ultrasonic senders and receivers, which generate a sound impulse in air and receive it after reflection from the surface of the object to be measured, are very suitable. The use of sound in air with its relatively low propagation velocity of $c_0 = 331.2 \, \text{m}^{-1}\text{s}$ under normal conditions ($T = 0°C$, $p = 101325 \, \text{N m}^{-2}$, $\rho = 1293 \, \text{kg m}^{-3}$) allows, in principle, a high resolution. Limitations to the available accuracy occur in the air through which the sound waves are propagated. The condition of a gas is described by the properties pressure $p$, density $\rho$ and temperature $T$. For propagation of sound in gases, the adiabatic law applies, namely that changes in pressure and density during the propagation of sound occur so rapidly that no heat exchange with the environment takes place:

$$c = \sqrt{K\frac{p}{\rho}} = \sqrt{\frac{E_v}{\rho}} \tag{65}$$

where $E_v$ is the bulk modulus. The adiabatic exponent $K$ is obtained from the adiabatic equation $pV_K = $ constant, and is temperature dependent. The density $\rho$ is relatively independent of pressure and humidity and is again mainly dependent on temperature. The velocity of sound in air relative to temperature can be expressed to a good approximation by the following equation:

$$c(T) = c_o \sqrt{\frac{T}{T_o}} \tag{66}$$

Here, $T_0 = 273 \, \text{K}$. This means that for a change in temperature of 20°C there is a change in the velocity of sound of 0.17% $\text{K}^{-1}$, which corresponds to $0.59 \, \text{m s}^{-1} \, \text{K}^{-1}$. Because of the relatively high temperature dependence, simultaneous measurement and incorporation of air temperature is necessary for accurate displacement measurements using ultrasound. If there is an error of 0.1 K in the measurement of air temperature, then there is an uncertainty in the measured value of $\pm 0.85 \times 10^{-5}$. At a distance of 200 mm this error would amount to $17 \, \mu\text{m}$.

The low propagation velocity of ultrasound relative to that of electromagnetic waves on the one hand gives easily measurable time differences because of the small differences in distance, but on the other limits the maximum measurement frequency. If equation (63) is solved for pulse frequency $f_p$, then we obtain

$$f_p = \frac{c}{2d_{max}} \tag{67}$$

If $c$, the velocity sound, is taken at about $330 \, \text{m s}^{-1}$ and $d_{max} = 1 \, \text{m}$, then a maximum pulse frequency of 115 Hz is obtained. So long as the distance to be measured does not alter, or at least only alters slowly, a low pulse frequency does not present a problem. However, the measurement of moving machine parts is severely limited by this effect. A further disadvantage of acoustic displacement measurement is the difficulty in bundling the transmitted sound beam. A sender with an accurate directional characteristics is essential for selective measurement of distance to a given point on a machine. A suitable directional characteristic is possible with a large antenna, divided into various individual elements. Each individual element can be activated for amplitude and phase [86]. This type of antenna has long been used in radar technology and is called a 'phased array'.

Undesired echoes can be filtered out using electronic filter circuits, if the approximate position of the desired signal is known and there is a sufficient time lag between information signal and noise signal.

In devices for checking rotation of crankshafts [85], a condenser microphone is used. These microphones consist of a metallized plastic foil which acts as a diaphragm and a dielectric. The sender is fixed to a cruciform surface, divided into concentric zones, which are independently activated. Eight temperature sensors are used for temperature compensation within the apparatus, their signals being passed to a microprocessor-controlled correction unit. Temperature measurement is carried out to an accuracy of 0.1°C. To obtain a higher sensitivity in displacement measurement, the shortest possible duration of sender pulse is necessary. This is because only time lags greater than the pulse duration can be measured. An ultrasonic displacement meter with a bandwidth $B$ has the following distance resolution $\Delta x$:

$$\Delta x > 0.5 \frac{c}{B} \qquad (68)$$

A large bandwidth can only be obtained with a correspondingly high carrier frequency. In the previous case an ultrasonic frequency of about 350 kHz was selected. Since the damping effect of air increases with frequency and thus reduces the distance which can be measured, an upper limit of 1 MHz is only possible at small distances ($\leqslant$500 mm). In relative measurements, errors of the order of one hundredth of a millimetre may occur. A particularly useful application of ultrasonic measurements is in level measurement [87;88]. The levels in tanks and silos for fluids and granular materials can be continually subject to contactless measurement. In practice, accuracies of between 1 mm and 2 cm are obtainable [88]. The use of reference surfaces permits calibration of the equipment and therefore higher accuracy. In most cases sender and receiver are combined. This is especially true of piezoelectric ultrasonic transmitters, which can either produce mechanical vibrations in response to electrical stimulation, or vice versa. Separation of these functions means lower expense on control and evaluation electronics. To obtain the lowest possible measurement error, sender and receiver must be set as closely as possible together, with their axes slightly tilted towards one another. Three examples of this are: a system with a range of 0.75–600 mm with a resolution of 0.25 mm; another system with a range of 600–9000 mm and a resolution of 10 mm; and a third for a range of 10–50 m with a resolution between 1 and 20 mm [81–84;88].

### 3.9.3   Laser distance measurement

For accurate measurement of the distance of point objects at ranges up to several kilometres, solid-state lasers are used because of their outstanding properties. The laser beam has a high spatial coherence, which, combined with the close bundling, permits closely adjacent measurement points. The spread of the laser beam is only of a few tenths of a milliradian. The beam is of high spectral purity and coherence, which makes for easy filtering and correction for noise and allows the use of heterodyne receivers (superposition receivers). The high intensity available with solid-state lasers, in which individual pulses may have a power of between 10 and 100 MW, allows their use over ranges up to 10 km, despite strong atmospheric absorption. There is also the capability of producing extremely short pulses in the nanosecond region, which allows good time lag measurement with high resolution [77]. Compared with electromagnetic instruments, lasers have the advantage that the refractive index of the normal atmosphere is only slightly dependent on temperature, pressure and humidity. Within the visible region the refractive index varies by as little as $10^{-5}$ [89].

The basic setup of a laser distance meter using the pulse-time process is shown in Figure 3.116. The laser sends a beam of light to the object to be measured which, depending on application, may have a triple reflector or, corresponding to its reflective properties, returns the beam more or less diffused over the half-distance. A silicon photocell or silicon avalanche photodiode is used as a receiver. The heart of the equipment is a computer whose start and stop commands indicate the time $t_0$ and $t_1$ of the sender or receiver pulse. The time difference $t_1 - t_0 = \Delta t$ determines the distance. The accurate displacement $d$ is given by the equation:

$$d = \frac{1}{2} c . \Delta t \tag{69}$$

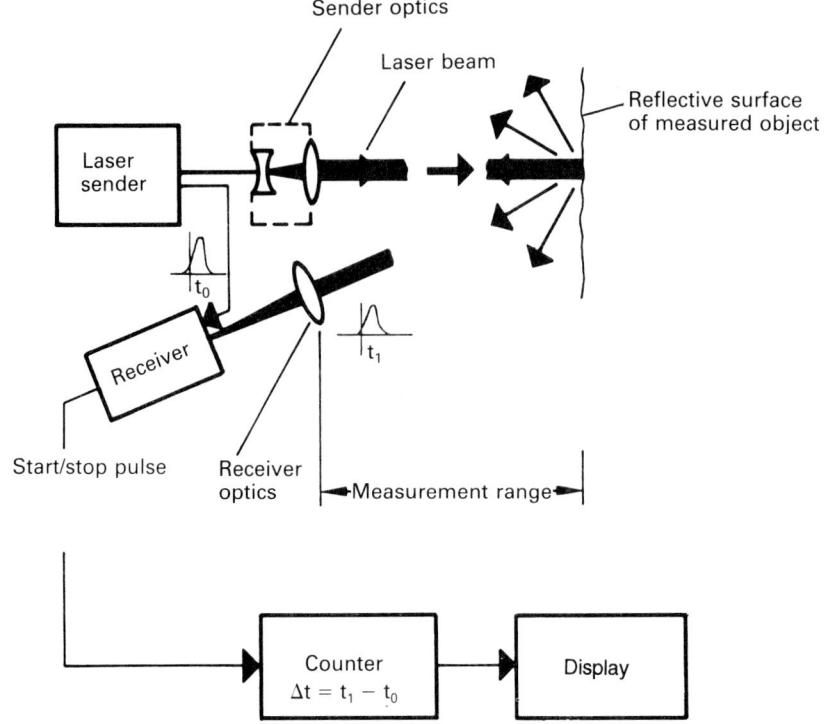

Figure 3.116   Laser distance meter.

where $c$ is the velocity of light ($2.997925 \times 10^8 \, \text{m s}^{-1}$). Materials for solid-state lasers include ruby ($\lambda = 0.6943 \, \mu\text{m}$), especially the YAG-neodymium type ($\lambda = 1.06 \, \mu\text{m}$), neodymium glass ($\lambda = 1.06 \, \mu\text{m}$) and GaAs ($\lambda = 0.86 \, \mu\text{m}$). For long range systems YAG-Nd and glass-Nd are especially good. Semi-conductor lasers (GaAs) are rather less powerful. Since higher pulse-follower frequencies are possible, this disadvantage can be partially compensated for on the receiver side by suitable integration processes, so that systems with GaAs lasers can be used over shorter ranges.

   The accuracy of the displacement measurement depends on the accuracy of the time measurement between the transmission of the primary pulse and the reception of the reflected beam. In this, the slope of the wavefront, the signal-to-noise ratio and the properties of the wave detector play a decisive role. These errors can add from 1 to 2 m to the errors resulting from counting a multiple vibration period of a quartz-stabilized oscillator. In a practical example [90], an error of ± 10 m occurs for a maximum range of 20 000 m. This corresponds to an error of ±0.05%. The use of a reflector fixed to the object to be measured, as in an American aircraft position measurement system, gives an error of ±0.3 m (1 foot) at a distance of 37 km (20 nautical miles). This corresponds to an accuracy of 0.00081%.

The pulse-follower frequency in laser distance measurement alters with the temperature of the laser from 1 to 20 Hz. Nevertheless, in many cases forced cooling of the excitation lamps and the laser rod itself is possible. The main application of laser distance measurement of the pulse-time measurement type is in the military field, where they are used in fire control systems in particular. For measurements with higher accuracy, phase measurement systems are used, with continuous-beam lasers. The laser beam acts as a carrier which is modulated by a frequency $f$, and the phase of the modulated vibrations is measured. The distance is given from the relative phase position according to the following equation:

$$a = \frac{1}{2}\left(\varphi \times \frac{\lambda}{2\pi} + n \times \lambda\right) \tag{70}$$

where $n$ is an integer, $\lambda$ is the wavelength of the modulation vibration and $\varphi$ is the measured phase angle. It is therefore an incremental process and not simple. This ambiguity can be removed by multiple measurements at different modulation frequencies [92, 93].

This technology is used in geodesy and also in sport. With interferometric processes, as described in the following section, accuracies in the region of $10^{-8}$ m are possible.

A laser rangefinder which can measure distances absolutely in the range 0–200 mm is described in [94]. The distance is determined from the intensity of the diffused reflected beam. Errors arising due to reflective measurement objects are eliminated through the use of polarized light and a suitable polarization filter on the receiver.


## 3.10   Laser interferometer

The laser interferometer forms an extremely accurate and highly sensitive measurement system for length standardization for workforce and workplace. It can therefore be used either to calibrate other measurement systems (e.g. machine tools and co-ordinate measuring machines) or for direct measurement on the machines themselves. Tilting and rotation of the machine axes in the direction of motion can be determined in order to avoid errors in processing. The laser interferometer is suitable not only for distance measurement, but also for the measurement of angles when used with suitable additional equipment. It can also be used for measuring flatness of machine beds or leads, for the determination of straightness of rails and for the parallelism or co-ordinate movement and the rectangularity of machine axes.

The main components are the laser, the interferometer and the electronics. Length measurement is achieved by splitting the light beam into two component beams, one of which travels along the distance to be measured while the other travels along a constant comparative length, and measuring the time lag between them. Since the absolute value of this time difference is extremely difficult to measure, the movement of a reference mark over the measured difference is measured to give the time difference as a relative factor. When the two beams are reunited they interfere with one another, such that the different phases either reinforce or cancel each other out in the photoelectric receiver, or even in the human eye. Figure 3.117 shows the principle of the Michelson Interferometer, which is one of the basic types of optical interferometer. A single beam splitter, which also reunites the beams, separates the beam into the measurement and comparison components, which are then reflected from two mirrors and reassembled and their interference measured. As the measurement mirror moves over the distance to be measured,

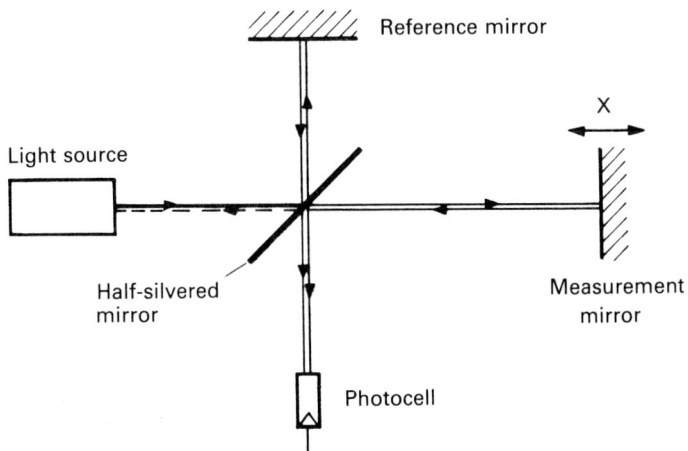

Figure 3.117　Operating principle of the Michelson interferometer.

the phase length of the measurement beam alters relative to that of the comparison beam. The photo-receiver records periodic light and dark bands, and the distance between two dark bands represents a displacement of the measurement mirror by half a wavelength of the light beam. So that the photo-receiver can record interference between both light beams, their wave fronts must run in parallel. If a plane mirror is used as the beam splitter, a minimal tilt disturbs the interference. In Figure 3.118, three types of mirror are shown, which allow tilts of up to a few degrees. The triple mirror is the most commonly used of these mirrors. It has the shape of the corner of a cube, so that each mirror surface makes an angle of 90° with the others. In highly accurate mirrors this angle is maintained to ±1 second of arc. With the lens and the concave mirrors, a mirror is placed at the focal plane of an optical system. Thus, all beams parallel to its main axis are reflected and an interference signal obtained through superimposition of two coherent partial beams is only a first approximation to the length measurement, since no determination of direction is possible. Because of this, the laser interferometer has two photo-receivers, of which one records the reference signal, usually phase-shifted through 90°. Laser and interferometer are often combined in a single unit, but this is not recommended. Simultaneous measurements of different distances using a single laser and several interferometers are possible when separate instruments are used. A practical example of this is the simultaneous measurement of all four axes of a machine tool.

Since the laser interferometer works incrementally, and initially delivers only numerical pulses, one or possibly two numbers are necessary for calibration. A computer is also necessary to correct for the effects of atmospheric temperature, pressure and relative humidity on the wavelength of the light. For a maximum error of $10^{-6}$,

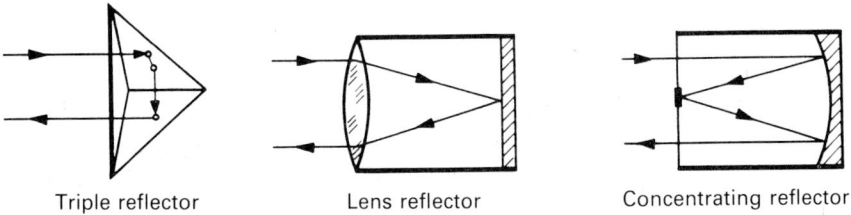

Triple reflector    Lens reflector    Concentrating reflector

Figure 3.118   Basic types of reflector. (a) Triple reflector. (b) Lens reflector. (c) Concentrating reflector.

atmospheric pressure must be measured to $400 \, \text{N m}^{-2}$, and temperature to $\pm 1 \, \text{K}$. If a distance of $10 \, \text{m}$ is to be measured to an accuracy of $\pm 1 \, \mu\text{m}$ (i.e. a maximum error of $10^{-7}$ is required), then air pressure must be measured to $40 \, \text{N m}^{-2}$ and temperature to $\pm 0.1 \, \text{K}$. Between 40 and 80%, the effect of atmospheric humidity can be neglected. With some laser interferometers there is a choice between manual input of data on ambient conditions and the use of a special sensor that inputs the corrections automatically. In addition to changes in the refractive index of air and its effect on the wavelength of the laser light, the temperature of the test piece itself plays a significant role. Commonly encountered materials in machine tool applications have coefficients of linear expansion around $10^{-5} \, \text{K}^{-1}$. This temperature coefficient can lead to much greater errors than changes in the wavelength of light due to changes in ambient conditions. For this reason, measurement of machine or workpiece temperature is much more important. Suitable sensors are available for this, which in the best case allow the measurement of workpiece length at normal temperature to be read directly off the laser interferometer. A final important feature of computers is that they permit readout in whatever units are desired—metres, millimetres, or inches.

An example of a laser interferometer which is suitable for both labour and workshop operations is given in [99]. Further data will be found in [38;40;100–106].

The complete measurement equipment consists of a laser interferometer, a triple mirror, evaluation electronics with computer and a printer for automatic registration of results. Extra evaluation and registration equipment can be connected via an IEEE 488 interface bus. Additional equipment for simpler operation includes sensors for recording the temperature of the machine or workpiece to be measured (the temperature coefficient of the material must be known). Sensors for pressure, temperature and relative humidity (atmospheric parameters) allow automatic correction for variations in the local velocity of light. The apparatus is ready for use after a warm-up period of about $10 \, \text{min}$. A special monomode He–Ne laser is used (wavelength in a vacuum $\lambda = 0.623991 \, \mu\text{m}$), and its light is passed through a built-in Zeeman cell to give one right and one left-hand circularly polarized wave with a frequency difference of $1.8 \, \text{MHz}$. This can therefore be described as a double-frequency laser [86]. The beam is spread via a telescope, and passes through a neutral beam splitter to give a reference beam, Figure 3.119. The other component of the beam is led to a modified Michelson interferometer. A polarizing filter in both measurement and comparator beams

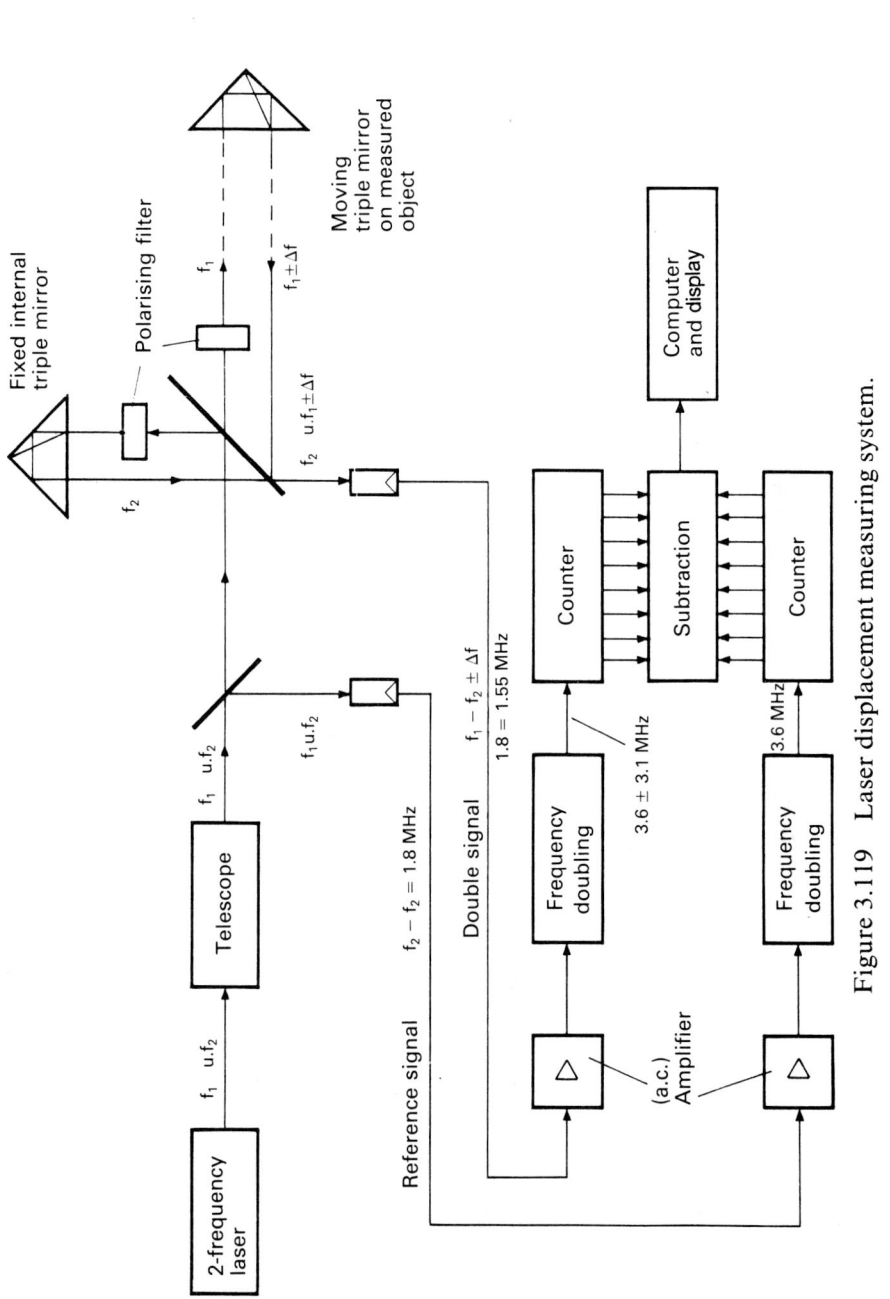

Figure 3.119  Laser displacement measuring system.

separates the two frequency components, so that the measurement section has a beam of frequency $f_1$ and the comparator section one of frequency $f_2$. After the beams are reunited, they pass on to a photosensitive cell. This produces a beat signal of frequency $f_1 - f_2 =$ 1.8 MHz. A change in length of the measurement beam due to movement of the triple mirror produces a change $\Delta f$ in its frequency $f_1$ because of the Doppler effect, up to a maximum of ±1.55 MHz. The output signal of the interferometer therefore lies within a range 1.8 ± 1.55 MHz. In contrast, the reference signal has a constant frequency of 1.8 MHz. Both signals are amplified via a.c. amplifiers and are passed on to two individual frequency counters after frequency doubling. One advantage of this process is that the use of d.c. amplifiers, with their known disadvantages relative to a.c. amplifiers, is avoided. The measured value is derived by subtraction of the two counter outputs. In this way the measurement direction is also given simultaneously. Each increment of the counters represents a quarter of a wavelength ($\lambda/4$). The resolution of the display is therefore 0.2 $\mu$m. By averaging several successive measurements (digital phase interpolation), a resolution of 0.01 $\mu$m can be obtained.

Accuracy depends on the type and accuracy of the determination of ambient parameters, and thus on the velocity of light. If this is accurately known, then errors as low as $\pm 0.1 \times 10^{-6}$ can be obtained. With automatic measurement of atmospheric parameters, error increases to $\pm 3 \times 10^{-6}$ (operational temperature 0–40°C). Maximum range for distance measurement is 40 m with a movable reflector. Velocity measurements up to 18 m min$^{-1}$ are possible with an error of ±0.1%. Resolution is therefore 0.1 mm min$^{-1}$.

## 3.11 Laser gyroscope

The possibility of using a ring-laser to measure angular movement was proposed in 1963, shortly after the discovery of the laser. The basic form of a laser gyroscope is shown in Figure 3.120. Three holes are drilled into a glass body in the form of an equilateral triangle. Mirrors are mounted at the apices such that a closed ring is formed for a light beam, which is induced in the cylindrical gas-filled bores by means of a laser effect. The laser light is produced in two gas-pressurized lengths with built-in cathodes. The three mirrors form the resonance space for two He–Ne lasers with opposed wave paths. The wavelength is 0.6328 $\mu$m ($f = 4.74 \times 10^{14}$ Hz). The resonant frequency of the laser is a function of the optical path length. Thus, both

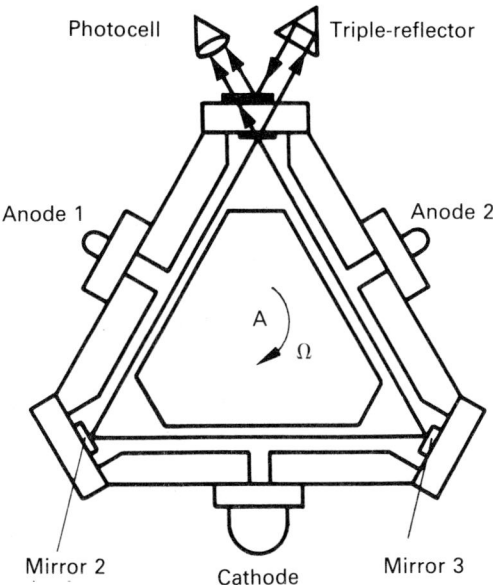

Figure 3.120   Laser optical gyroscope.

laser beams have identical frequencies when the system is at rest. If, however, the laser gyroscope experiences a rotary movement perpendicular to surface $A$, in which the two beams move, then the path lengths travelled by the two beams will differ. The beam which is propagated in the direction of motion must travel a longer path than the other beam which moves against the direction of motion. Thus, the resonant frequencies of the two lasers are altered in proportion to the angular velocity $\Omega$ [107;113]:

$$\Delta f = \frac{4A\Omega}{\lambda L} \tag{71}$$

Here, $A$ is the surface area enclosed by the laser beams, $L$ is the resonator length, $\lambda$ and $\Omega$ the angular velocity.

Mirror 1 allows a small percentage of the laser light to escape, so that the two beams, as shown in Figure 3.120, can be made to interfere. When the system is rotated, the interference fringes move and the photodetector records alternating light and dark bands, which can then be counted. Thus the angle of rotation can be determined by simply counting the interference fringes.

In the practical use of laser gyroscopes it has been shown that between the two opposite modes, through unavoidable stray reflection at the mirrors, energy is interchanged, which for small rates of rotation (up to about $500°\,h^{-1}$) can lead to a synchronization of the wave trains ('lock-in effect') [109].

This effect can be compared with electrically coupled systems of oscillators. To avoid this synchronization effect there is the possibility of setting the whole system into mechanical vibration, or fixing of one of the resonator mirrors to a piezoelectric crystal, which itself can be set into vibratory motion. Thus, both laser systems are set into a forced coupling.

The laser gyroscope is used in large modern aircraft as a sensor for rotary motions and angles. Because of its highly technical nature and the manufacturing costs associated with it, its use is mainly limited to aerospace and military technology. An alternative is the fibre gyroscope, in which the light path is a length of monomode optical fibre and the light source a GaAs diode, Figure 3.121. The basic principle is very simple. One of the light bundles leaving the single light source is split into two by means of a half-silvered mirror, which is arranged to be at opposite ends of a coiled optical fibre. By means of the same mirror, the two partial beams are made to interfere after passing through the optical fibre. If the coil is rotated the two beams pass along paths of different lengths, giving rise to a phase shift $\Delta\phi$ (the Sagnac effect).

This is:

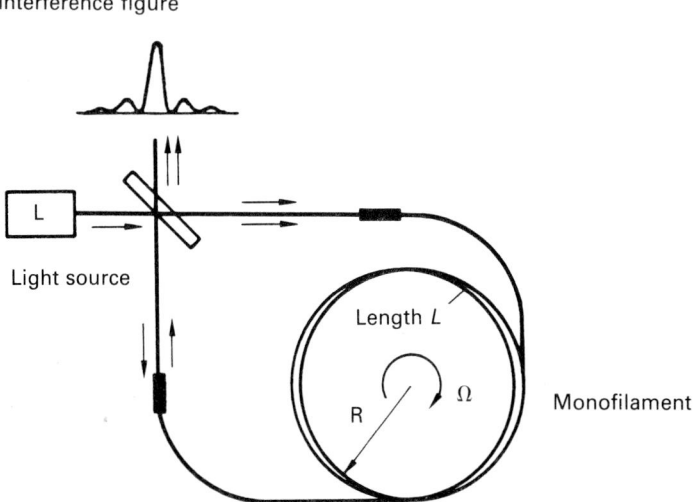

Figure 3.121    Fibre optical gyroscope with single mode fibre.

$$\Delta\phi = \frac{4\pi LR}{\lambda c}\Omega \qquad (72)$$

where $L$ is the length of the fibre, $R$ the coil radius, $\lambda$, the wavelength, $c$ the velocity of light and $\Omega$ the angular velocity. The effect to be measured is very small. For an angular velocity of $1°\,h^{-1}$, for example, a time difference of $\Delta\phi = 10^{-20}$ or phase differences of $\Delta\phi = 10^{-20}$ must be measured. This is only possible with a range of additional measures, which we cannot go into here. As a test, a fibre of 0.5 km in length was made into a coil of 12 cm diameter (1325 windings). The core diameter of the fibre was $4\,\mu$m, the external diameter $100\,\mu$m, while a semi-conductor diode laser ($\lambda = 850$ nm) was used as a light source, and a p–i–n photo-diode was used as a sensor.

The minimum angular velocity measurable with this equipment is around $3°\,h^{-1}$ [108;111;112]. Current developments are concentrating on a further increase in the useful range and thus of a general miniaturization using integrated optics. It is expected that in the near future an instrument will exist which will show advantages over the ring-laser: this will occur with the use of long-life solid-state components that are also used in signal technology and are thus available at reasonable prices. The ageing rate of the laser diodes used is so small that a service life of 100000 h may be expected. Also, high voltages are not required. The simple technology will therefore make low prices possible. Laser gyroscopes and fibre gyroscopes were originally developed and designed for applications in navigational equipment in aircraft, ships, vehicles and missiles. A new and very important application will be in industrial robots. In contrast to the more usual angle measurement systems, which measure an angle between two contrary moving machine parts and are thus relative, in this case an absolute angle can be measured relative to a fixed basics co-ordinate system. The significance of this characteristic will be clear in permitting measurements of the movements of human hands and arms up to the shoulder joint. If a particular arm movement and thus a movement of the tip of the index finger is required, then simultaneous rotary movements of the finger joints, the hand joints, the elbow joint and the shoulder joint are possible. If it is required to reproduce this mechanism in a robot arm, then a total of six angle measuring systems for the determination of one angular movement of the last joint of the index finger would be necessary. With conventional sensor and control designs, the obtainable positional accuracy of the measurement technology employed is limited by the fact that variations in the measurement of the angular position of one limb are fully reproduced in the resulting calculation of absolute angular position of all successive limbs. A deviation in the measurement of the angle of the shoulder joint thus affects the calculation of the elbow joint, and so forth. With six joints, therefore, considerable angular errors, and hence positional errors, can arise. With the use of angle sensors, the angle is measured absolutely relative to a basic co-ordinate system, and so any errors are confined to the individual co-ordinates of one arm and do not affect other angular errors.

The development of optical absolute angle sensors opens up technology which has previously been limited to aerospace to general industrial use. This development is by no means at an end. The possibilities of modern micro-optics and semiconductor physics in this area are not by any means exhausted. In addition to robot technology, there are many other applications of rotary angular measurement in industry, where no extremely high rate of rotation occurs. The development of simple semiconductor laser gyroscopes on the basis of currently available laser diodes provides new solutions for areas of production measurement technology, especially in the measurement and inspection of precision machinery, co-ordinate measurement equipment and production equipment for large vehicles, aircraft or ships. It is assumed that the following specifications will be reached: angular resolution of approximately $1 \mu$rad, rate of rotation, $<1 \mu$rad s$^{-1}$, and the possibility of compensation for the rotation of the earth. Currently realisable accuracy (angular resolution approximately $10 \mu$rad, rate of rotation $100 \mu$rad s$^{-1}$, drift rate approximately $0.1$–$1$ mrad h$^{-1}$) is however quite sufficient for many applications [109].

## 3.12 Systems with analogue measurement and subsequent analogue-to-digital conversion

In addition to simple digital measurement, there is a group of analogue measurement sensors which can be used in combination with analogue-to-digital converters and can play a significant role in measurement technology. A characteristic of most of these systems is the fact that an analogue range which gives an absolute signal must be periodically replaced. For this reason such systems are often designated as cyclic. Since, in fact, they deal with incremental processes with interpolation of individual increments, this description is misleading. In many systems the measured value is expressed as a ratio of two sinusoidal voltages, of which one is amplitude modulated proportionally to the sine and the other proportionally to the cosine of the measured magnitude $x$:

$$u_2 = L\hat{u}_1 \sin\omega t \cos x \qquad (73)$$

$$u_3 = K\hat{u}_1 \sin\omega t \sin x \qquad (74)$$

Despite external differences in construction and method of operation of the sensor, similar principles and processes for the digitization of the measured values are usable.

### 3.12.1 Distance measurement with differential capacitor

An application of the principle of the differential capacitor for distance measurement with a resolution of $1\,\mu\text{m}$ and a measurement range from 25 mm to 2000 mm is described in [92]. The basic principle is that a parallel displacement of a plane electrode relative to two electrodes of the same potential set opposite to it changes its partial capacity in a linear relationship to the displacement $x$, Figure 3.122. Changes in the distance between the electrodes or the dielectric constant have only a minor influence. The sum of the partial capacities $C_{\text{ges}} = C_{13} + C_{23}$ is constant in the hatched area, in which no boundary effects exist. The ratio of one of the two partial capacities to the

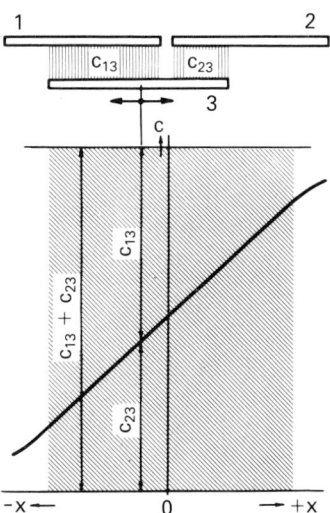

Figure 3.122  Curve of partial capacitance of differential capacitor when electrode 3 is displaced. The usable linear section is shaded.

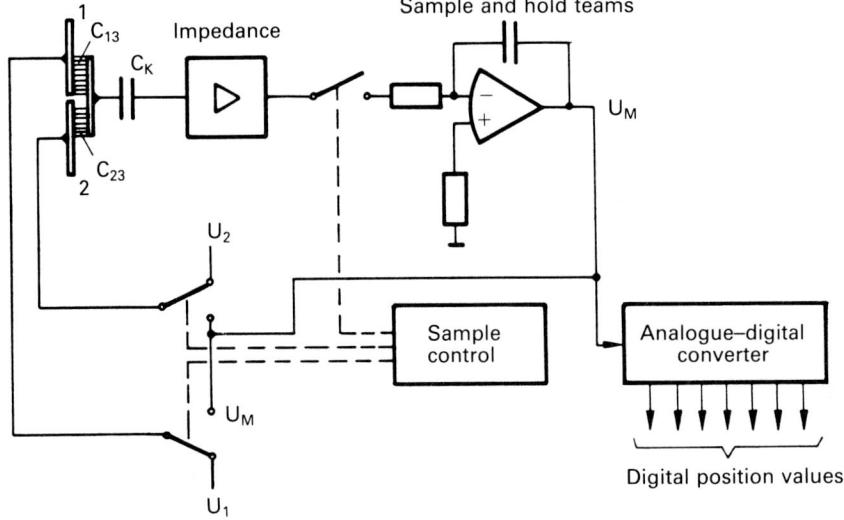

Figure 3.123   Servo loop for converting a capacitance ratio into a d.c. voltage $U_M$.

total capacity changes with respect to the displacement $x$ in an almost linear relationship:

$$x \sim \frac{C_{23}}{C_{12} + C_{23}} \tag{75}$$

To measure the ratio $C_{23}/(C_{12} + C_{23})$, a null-seeking circuit may be used, as shown in Figure 3.123. Here, the partial capacities $C_{13}$ and $C_{23}$ are periodically charged, so that the charges $Q_{13}$ and $Q_{23}$ are compensated for by a suitable selection of $U_M$ and electrode 3 remains at constant potential so long as there is no alteration of the partial capacities as a result of a displacement of electrode 3. The voltage on electrodes 1 and 2 is shown in Figure 3.124. The alternating voltages on $C_{13}$ and $C_{23}$ cause a vibration process that is not affected by the closed null-seeking circuit, so long as the pick-up of $U_3$ follows its decay.

Assuming similarity, then:

$$Q_{13} = Q_{23}$$

from which:

$$(U_M - U_1)C_{13} = -(U_M - U_2)C_{23}$$

is obtained. If this is solved for $U_M$, then we obtain:

$$U_M = U_1 + (U_2 - U_1)\frac{C_{13}}{C_{13} + C_{23}} \tag{76}$$

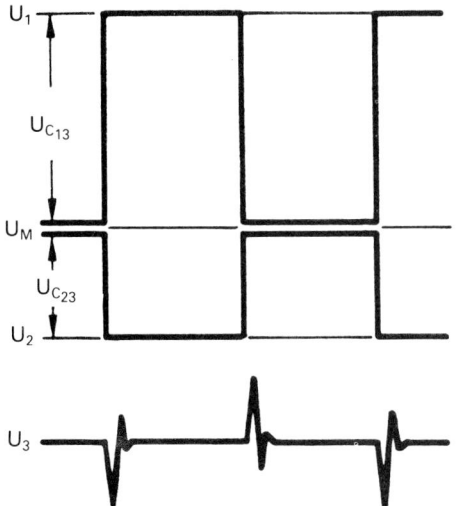

Figure 3.124   Gradient of voltages at electrodes 1, 2 and 3.

Since $U_1$ and $U_2$ are constant voltages (e.g. 10 V and $-10$ V), $U_M$ is proportional to $C_{23}/(C_{13} + C_{23})$ and is thus directly proportional to displacement $x$.

The task of the null-seeking circuit is so to control the measurement voltage $U_M$ that the above condition is fulfilled.

### 3.12.1.1   Measurement of arbitrary lengths with switchable electrode elements

With a differential capacitor, the technically feasible linearity which may be obtained limits the accuracy of the measurement range to a few millimetres. This limitation may be circumvented if both electrodes 1 and 2 are made up of several parallel-switched strips, and the arrangement of the strips during relative motion between electrode 3 and the two other electrodes changes when a given minimum value of the measurement voltage $U_M$ is attained, as shown in Figure 3.125. The strips in electrode 1 lying behind the direction of motion are switched and therefore the last strip in electrode 2 is exchanged for that of electrode 1. Electrode 2 therefore contains a new strip in the direction of motion. Thus, electrodes 1 and 2 follow the motion.

In practice, electrodes 1 and 2 on the moving measurement head and electrode 3 are in a comb-like arrangement with one another,

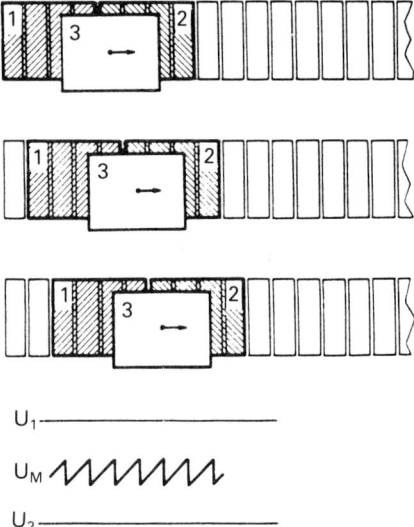

Figure 3.125   Basic construction for various measurement intervals.
Electrodes 1 and 2 of the above differential capacitors are formed by
interconnecting conductive strips. When the strips are switched over the
electrodes thus formed 'follow' the movement of electrode 3. The switch-over
becomes automatic if the voltage $U_M$ exceeds a specific limit value.

Figure 3.126a,b. In the intermediate space between the individual
strips of the comb, the average displacement of which is 4 mm, there
is an earthed protector. All strips of the scale make up electrode 3,
where each is only partly electrically operative, since the readout
head and its controlled electrodes 1 and 2 are set opposite one
another. A peculiarity of electrode 3 is that it is only capacitatively
coupled to the input amplifier of the tracking system. The opposite
electrode of the coupling capacitor is located on the readout head in a
plane with electrodes 1 and 2. A weakening of the signal of electrode
3 through capacitative scanning is not a problem, since it only serves
as a criterion for the comparison of the tracking system. It is,
however, important to avoid direct coupling of the pick-up electrode
to electrodes 1 and 2 through the steel scale. The desired indirect
coupling through electrode 3 can be obtained by insulation and
through mounting the impedance converter directly on the rear side
of the coupled electrode.

From Figure 3.126 it can be seen that this is made up of approximately 64 conductive strips 0.5 mm wide, of which four are connected
to electrode 1 and the next four to electrode 2. Eight adjacent strips,

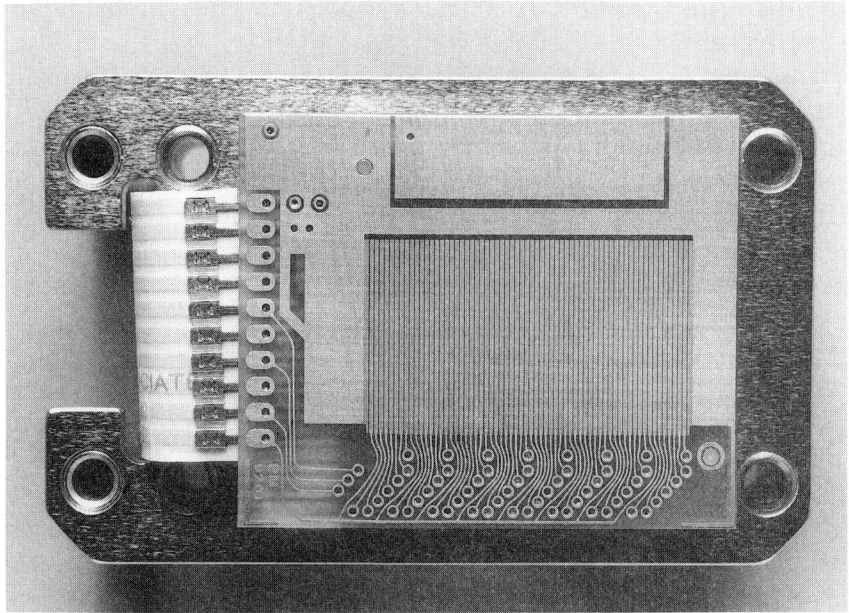

Figure 3.126a    Read head. (Picture courtesy of SYLVAC-Métrologie.)

Figure 3.126b    Scale. (Picture courtesy of SYLVAC-Métrologie.)

together with one of the 4 mm wide strips of the scale, form a differential capacitor.

Since all electrical connections are on the moving measurement head, the system is protected from external errors. Through the simultaneous readout of several successive differential capacitors consisting of eight-electrode groups of the measurement head, the effect of component errors is also reduced. The production of scale and measurement head resembles the Inductosyn scale described in Section 3.13, with copper electrodes etched on an insulated intermediate length of epoxy base on a steel body.

At the moment, scale lengths of up to 2000 mm are in use, giving a resolution of $1 \mu m$ and a measurement error of $5 \mu m$ on a 500 mm measurement length.

The measurement processes described so far are expressed not as a result of pure analogue measurements but as a combination of analogue and digital technology. A pure analogue measurement is possible with a coaxial arrangement as shown in Figure 3.127. Here, an earthed metal screen is inserted into an arrangement of two cylindrical capacitors. The first capacitor $C_R$ serves as a reference and the second $C_M$ as a measurement capacitor. The screen electrode experience as a change in capacity $C_M$ strongly proportional to displacement $x$ if the inner and outer electrodes of the measuring capacitor are sufficiently cylindrical. The measurement of the ratio of $C_M$ to $C_R$ is carried out according to the principle of charge comparison. The capacitors are charged with square-wave voltages of

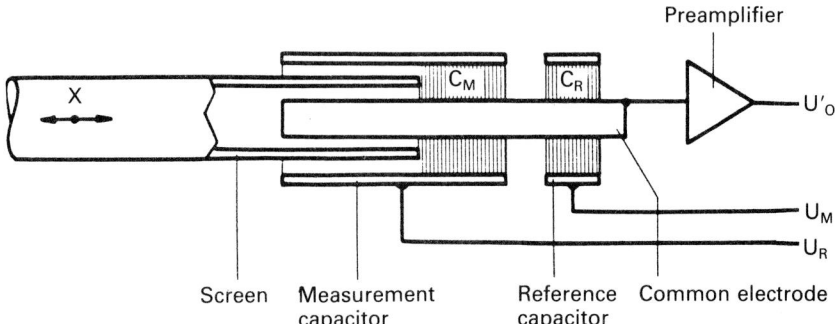

Figure 3.127   Schematic construction of a measuring probe. The capacitance $C_M$ of the measuring capacitor is altered proportionally to the displacement by introducing a conductive screen. The reference capacitor compensates for environmental influences. A preamplifier is connected to the load side of the common electrode in order to avoid external coupling.

alternating polarity and then subsequently discharged, from which the measuring capacitor is controlled with a voltage $U_M$ with constant amplitude. The amplitude of the reference voltage $U_R$ is thus controlled (null-seeking circuit) until the whole electrode experiences no further charge, i.e.:

$$Q_n = -Q_R$$

or:

$$U_M C_M = -U_R C_R$$

then:

$$U_R = U_M \frac{C_M}{C_R} \tag{77}$$

and is thus directly proportional to the displacement $x$ of the screen.

For the layout described, the maximum deviation in linearity is $1\,\mu$m over a measurement length of 25 mm [114].

## 3.12.2  Resolvers

A resolver is in principle a rotary transformer consisting of a rotor and a stator. The stator, and in most cases the rotor, consists of coils which are set at 90° to one another, Fig. 3.128. In its basic construction, the resolver resembles a very small ac generator. Because of its robust design, its small size and its high accuracy, it has wide application in the field of measurement technology. Numerous processes for

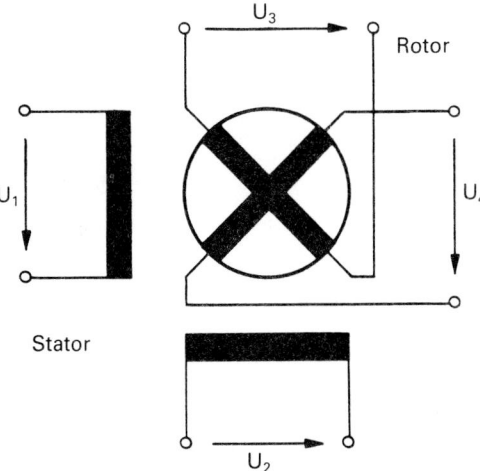

Figure 3.128   Winding diagram of resolver.

digitizing its analogue output are available within the field of data processing. Apart from the resolver, there are numerous other inductive devices for measuring angles, which in general are known as synchros. Although operating on the same basic principles, they may

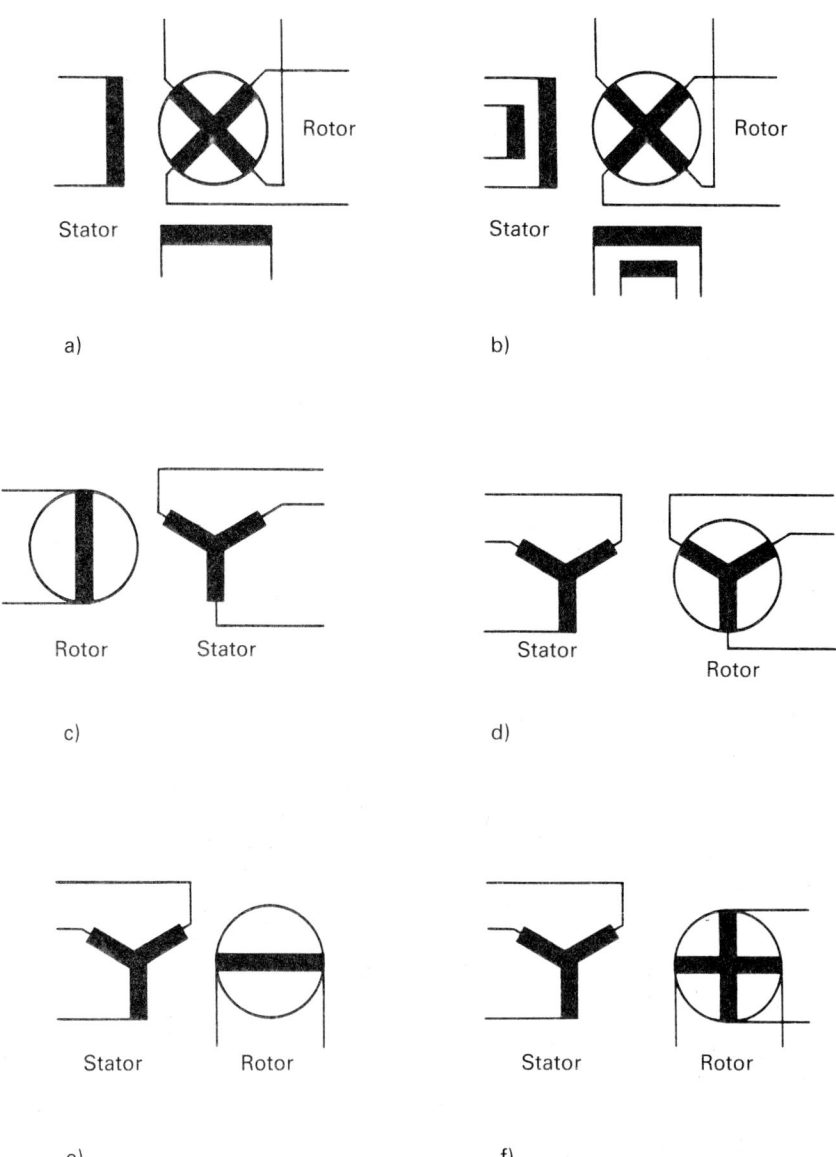

Figure 3.129   Designs of resolvers and synchronisers. (a) Resolver; (b) Resolver with compensating windings. (c) Control transmitter. (d) Control differential transmitter. (e) Control transformer. (f) Transolver.

be distinguished from one another by the type and combination of their coils, Figure 3.129 [115–118].

### 3.12.2.1 Design and function

The basic principle of these devices is the varying mutual inductance between the primary and secondary windings. In the case of a resolver, the definition of a winding as primary or secondary is entirely arbitrary. It is preferable, therefore, to base the distinction on a definition of rotor and stator windings. The rotor windings are usually led to a slip-ring. There are, however, also arrangements in which the windings are led to a ring-transformer. The mechanical construction of a resolver is characterized by extreme requirements on material properties and mechanical processing. The stator is made up of single slotted steel sheet and carries a double-core coil. In most systems the stator slots are chamfered in order to eliminate angle errors and magnetic retention forces. The cylindrical rotor has slots containing pairs of coils opposed by 90°. Figure 3.130 shows the form of the rotor and stator sheets. The rotor runs in prestressed roller bearings which are therefore free of play. The typical rotor–stator air gap is 50 μm. To achieve this, the stator rod must be ground circular to an accuracy of ±1 μm. In the same way, the rotor is ground and polished in order to obtain the necessary circularity. The selection and processing of the magnetic sheet and the manufacture of the windings likewise requires great care. Only in this way can

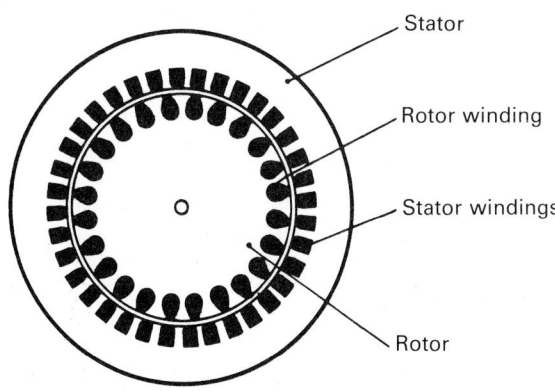

Figure 3.130   Design of rotor and stator cores for resolver.

Figure 3.131   Resolver (size 11).

the accuracy of these analogue sensors be guaranteed. Figure 3.131 shows an example of the usual form of a resolver, which is now thoroughly standardized through extensive use. They are usually provided in sizes based on diameter. There are sizes 5–28, which represent the diameter in tenths of an inch. The most important electrical property of the resolver is its ability to transmit voltage, which is a function of the mechanical angle between stator and rotor windings. It is defined by the pair of equations below, which are valid for a resolver with two 90° opposed stator and rotor windings:

$$u_3 = K_{13}\hat{u}_1 \sin(\omega t + \varphi_{13})\cos a + K_{23}\hat{u}_2 \sin(\omega t + \varphi_{23})\sin a \qquad (78)$$

$$u_4 = K_{24}\hat{u}_2 \sin(\omega t + \varphi_{24})\cos a + K_{14}\hat{u}_1 \sin(\omega t + \varphi_{14})\sin a \qquad (79)$$

In these equations $K_{ij}$ is the off-load transmission ratio between windings $i$ and $j$, $\varphi_{ij}$ is the phase difference between $u_i$ and $u_j$, $a$ is the mechanical angle between stator and rotor with reference to the magnetic axis of coils 1 and 3.

It is usually possible to simplify the above equations, since the transmission ratios between the four possible combinations of windings are the same:

$$K_{13} = K_{23} = K_{24} = K_{14} = K$$

Similarly, for the phase differences:

$$\varphi_{13} = \varphi_{23} = \varphi_{24} = \varphi_{14} = \varphi$$

$$u_3 = K(\hat{u}_1 \cos\alpha + \hat{u}_2 \sin\alpha)\sin(\omega t + \varphi) \tag{80}$$

$$u_4 = K(\hat{u}_2 \cos\alpha - \hat{u}_1 \sin\alpha)\sin(\omega t + \varphi) \tag{81}$$

Depending on requirements, therefore, a resolver can serve as an electromechanical modulator, as a function sensor, as a computer module or as an angle sensor. In most cases, the interest is mainly in its property of enabling angles to be measured to a high degree of accuracy. Here, only one of the two rotor windings is necessary. Depending on the type of connection, the process is therefore as described in equation (80) or, when the rotor winding is fed and the signal read off from the stator winding, is described by the two following equations:

$$u_1 = K\hat{u}_3 \sin(\omega t + \varphi)\cos\alpha \tag{82}$$

$$u_2 = K\hat{u}_3 \sin(\omega t + \varphi)\sin\alpha \tag{83}$$

These two equations correspond to equations (73) and (74). Figure 3.132 shows a line diagram of voltage $u_3$ according to equation (73) during rotation of the rotor.

Equations (73) and (74) or (82) and (83) apply for the transformed voltages, i.e. for the electrical no-load condition of the outputs for a rotor which is stationary or slowly rotating relative to the frequency of the input voltage. For this case, the electrical properties can be represented by a diagram which corresponds to that of a transformer

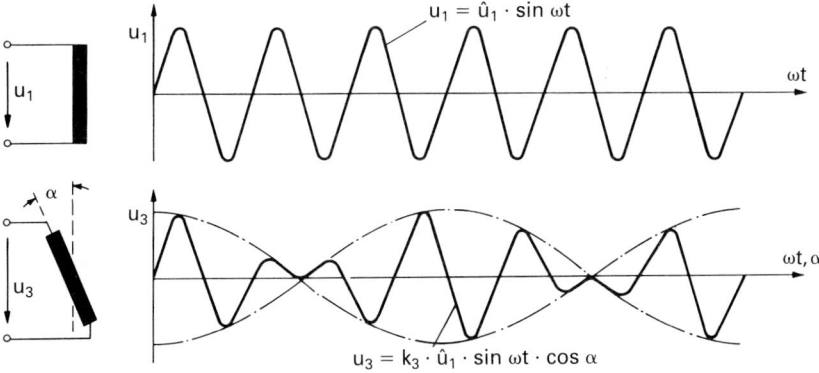

Figure 3.132   Line diagram of voltage $u_1$ and $u_3$ during rotation of rotor. $u_1$, Inductor voltage; $u_2$ output voltage.

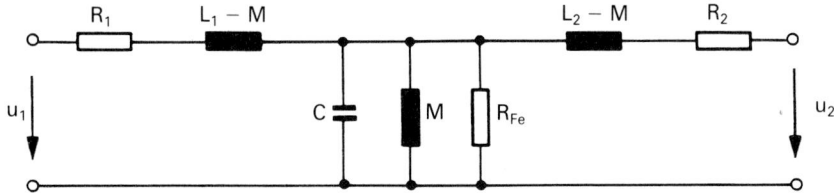

Figure 3.133   Equivalent circuit of transformer in resolver. $u_1$, Inductor voltage; $U_2$, output voltage; $R_1$, copper resistor in primary winding; $L_1$, inductance of primary winding; $C_1$, coupling capacitance; $M$, mutual inductance; $R_{Fe}$, equivalent core loss resistance.

with small eddy current losses, Figure 3.133, in which the mutual inductivity $M$ changes with rotation of the rotor. Depending on construction, there are different frequency responses for the voltage transmission ratio. There are therefore designs for fixed working frequencies, such as 50, 60 and 400 Hz and 10 kHz, as well as what are called wide-band designs with a frequency range of 25–200 kHz. The voltage transmission ratio $K$ usually lies between 0.25 and 2. In most cases it is 1.

The ratio $K$ is defined by:

$$K = \frac{\hat{u}_3}{\hat{u}_1} \tag{84}$$

The most important property of the resolver is its accuracy. In determining errors it is important to distinguish between static and dynamic errors. Static errors, which usually lie between the second and fourth harmonics in the error curve, are due to ovality of the rotor and asymmetry of the windings. Dynamic errors arise from rotation of the rotor, since voltage induced by the rotation are generated in addition to the desired transmitted voltage [119]. For the stationary condition, the voltage induced in a rotor winding by a single stator winding is:

$$u_3 = K'\omega t \cos\alpha \tag{85}$$

Rotation through $\Omega$ gives:

$$u_3 = K'\omega[\sin\omega t \cos(\Omega t + a_o) + \frac{\Omega}{\omega} \cos\omega \, t\sin(\Omega t + a_o)] \tag{86}$$

The second term of the equation is not necessary, and may be neglected when $\Omega \ll 1$. A further source of error lies in the no-load voltages. These are upper harmonic waves, generated by the curved steady-state characteristic of the core, as well as a fundamental signal 90° phase-shifted from the desired output signal, which arises through production tolerances in the manufacture of the windings. The phase rotation between input signals, which is caused by the resistance and inductivity of the windings and, depending on type of resolver, lies between 10 and 50°, is usually not important. A standard value for the maximum static angular deflection may be obtained from equations (82) and (83) as about $\Delta\varphi = 10$ minutes of arc. In precision resolvers errors of as low as 20 arc seconds are possible, and this value can be even lower when multipole resolvers are used.

### 3.12.2.2  Operating modes

For a resolver with two stator windings and one rotor winding there are in general three types of drive which have applications for digital angle measurement, see Table 3.8 [120;121].

- Feed both stator windings with in-phase sinusoidal voltages of different amplitudes—signal take-off from rotor winding. This process is known as amplitude control. A variant is often used

Table 3.8  Position measurement and control using a resolver

| Switching | Basic equation input | Basic equation output | Application |
|---|---|---|---|
| | $u_1 = \hat{u} \sin \omega t \sin \varphi$ <br> $u_2 = \hat{u} \sin \omega t \cos \varphi$ | $u_3 = K_3\, \hat{u} \sin \omega t \sin (\varphi - a)$ | Position measurement (ASV) |
| | $u_1 = \hat{u} \sin \omega t \sin \varphi$ <br> $u_2 = \hat{u} \sin \omega t \cos \varphi$ | $a = \arctan \dfrac{u_1}{u_2}$ | Position control |
| | $u_1 = \hat{u} \sin \omega t$ <br> $u_2 = \hat{u} \cos \omega t$ | $u_3 = k_3\, \hat{u} \sin (\omega t + a)$ | Position measurement (PSV) |
| | $u_1 = \hat{u} \sin \omega t$ | $u_2 = k_2\, \hat{u} \sin (\omega t \sin + \varphi_2) \sin a$ <br> $u_3 = k_3\, \hat{u} \sin (\omega t \sin + \varphi_3) \cos a$ | Position measurement |

for position measurement, when for a given fixed amplitude ratio the position of the rotor is controlled, up to $u_3 = 0$.

- Feed both stator windings with sinusoidal 90° phase-shifted voltages of the same amplitude signal take-off from rotor winding. Because of the phase displacement this is known as phase control.
- Feed the rotor winding—signal take-off from stator winding.

For each of these types of drive there is a method of obtaining digital readout.

### 3.12.2.3 Digital angle measurement using resolvers

#### Digital converter for resolver signals

The switching arrangements for converting resolver signals to digital values are usually known as resolver digital converters or synchro digital converters. In either case they are essentially analogue-to-digital converters. In contrast to many other measured values, angular information does not involve either the magnitude or the temporal characteristic of an individual voltage, but ratio of two amplitudes or two phase shifts. Since in both cases the voltages concerned vary with the sine or cosine of the angle of rotation, and their ratios must therefore be proportional to the tangent or cotangent of the angle of rotation, any switching arrangement for digitizing resolver signals must in addition to a conventional analogue-to-digital converter also have a component for the generation or evaluation of trigonometrical functions. Drive types 1 and 2 should first be compared, because of the similarity of the feed of two stator windings and signal take-off from a single rotor winding. In drive type 1 the angle information is given by the amplitude ratio of the input voltages, while in drive type 2 it is given by the phase difference between the output signal and one of the input signals (the reference signal). The essential distinction between these two processes lies in the great difference between their dependence on phase displacements between the two input signals. With amplitude control (drive type 1) the output signal $u_3 = K_3 \, \hat{u} \, \sin\tilde{\omega} t \, \sin(\varphi - a)$ is equal to zero when the electrical angle $\varphi$ and the mechanical angle $a$ are equal. With the position measurement, $\varphi$ is automatically controlled so long as a balance is retained. In a similar way, $a$ can be controlled by a null-seeking circuit for a given angle $\varphi$. An important criterion for comparison of the process is their liability to failure. For this reason the dependence of measured accuracy on amplitude errors of the feed voltage is first calculated. Thus the balance condition is related to the input signals $u_1$ and $u_2$. For amplitude control we have:

$$u_3 = K(u_1 \cos a - u_2 \sin a) = 0;$$

which leads to the condition $\tan a = u_1/u_2$.

For small deviations $\Delta(u_1/u_2)$, angle $a$ changes by $\Delta a$. This gives the condition $\tan(a + \Delta a) \approx \tan a + \Delta a$; thus $\Delta a = \Delta(u_1/u_2)$. If $\Delta(u_1/u_2) = 0.001$, then an angular error of $\Delta a = 0.057°$ is obtained.

The effect of phase displacement between the resolver signals, Figure 3.134, can be determined from the in-phase components of the two voltages.

The values $u_1$ and $u_2$ are the required signals for zero phase difference. In place of $u_2$ we now have $u'_2$, with a phase difference $\delta$ from $u_2$. The component of $u'_2$ in-phase with $u_2$ is $u'_2 \cos \delta$. An error of 1° per cent in ratio $u_1/u_2$ arises when $\cos \delta = 0.999$; this corresponds to an angle $\delta$ of 2.5°. If we now consider phase control (drive type 2) in terms of amplitude and phase errors in the input signal, it can first be established that there is no difference in amplitude deviations compared with the amplitude control process. The effect of phase difference is considerably greater.

For:

$$u_1 = \hat{u} \sin \omega t,$$
$$u_2 = \hat{u} \cos \omega t$$

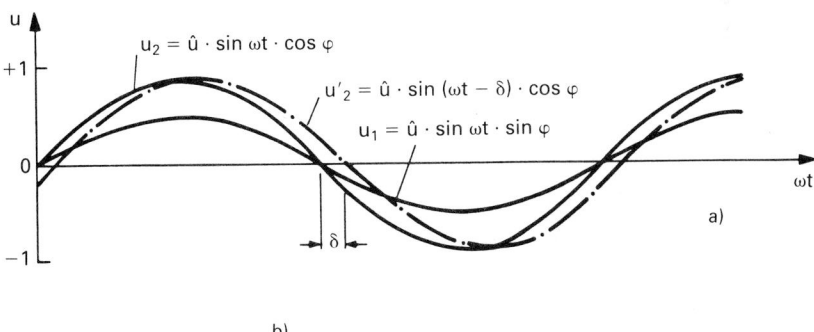

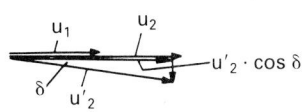

Figure 3.134   Influence of phase errors in amplitude control methods. (a) Line diagram; (b) Vector diagram. $u_1$, Inductor voltage 1; $u_2$, inductor voltage 2; $u'_2$, voltage phase shifted in relation to $u_2$; $\delta$, phase angle between $U_2$ and $U'_2$.

and

$$u_3 = K \sqrt{u_1^2 + u_2^2} \, \sin(\omega t + \varphi + a) \tag{87}$$

where

$$\varphi = \arctan(u_1/u_2)$$

If there is now a phase difference between $u_1$ and $u_2$, which varies from 90°, then there is also a change in angle $\varphi$, which is necessary for determining angle $a$. In Figure 3.135, $u_2'$ has a phase displacement of $90° - \delta$ from $u_1$. If $u_2'$ is then divided into two in-phase components $u_1$ and $u_2$, then we have:

$$u_2^* = u_2' \cos\delta$$

and

$$u_1^* = u_2' \sin\delta + u_1$$

If it is further assumed that $\Delta (u_1/u_2) = 0.001$, and it is required to obtain the phase difference $\delta$ associated with this, then the component of $u_2'$ in the sine sense is used, since the sine of a small angle varies very much more than its cosine.

$$\Delta(u_1/u_2) = \sin\delta \approx 0.001.$$

This gives $\delta = 0.06°$.

Comparing the amplitude and phase control processes, therefore, we find that with amplitude control there is a permissible phase shift of 2.5° for an error of 1° per cent, while with phase control the permissible phase shift is only 0.06°. Phase control is therefore more sensitive by a factor of 40 relative to phase shift of the input signal, which is very significant in terms of amplifier and filter design, as well as the requirements placed on the resolver.

If drive type 3 is used in a resolver, then only a single sinusoidal feed voltage is needed, in which merely a frequency constant adequate to the condition is set. Since the angle data is contained in the ratio of the two output voltages, variations in the amplitude of the input voltage affect only the sensitivity of the system. Amplitude control of the input voltage is therefore superfluous. The two secondary signals are amplitude modulated proportionally to the sine or cosine of the angle of rotation $a$. Their ratio is thus the tangent of $a$. There are conditions in which these signals are directly linked over phase-sensitive demodulators and are reprocessed over a common channel. With MOS technology a controlled switch can be manufactured to

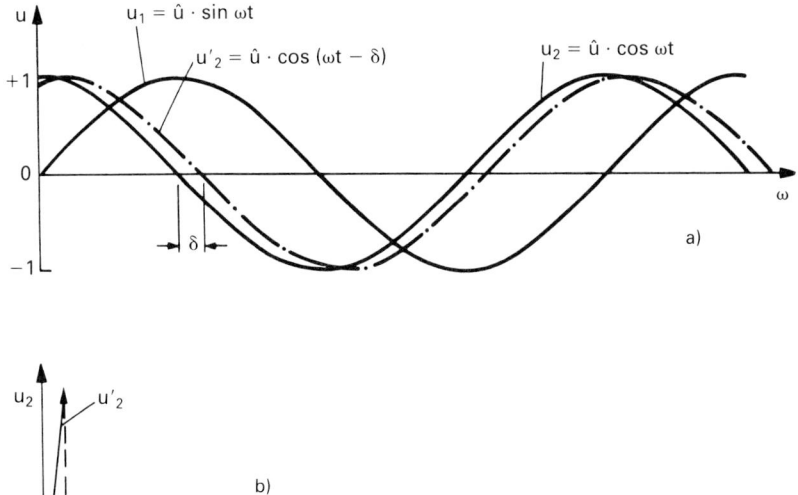

Figure 3.135  Influence of phase errors in phase control methods. (a) Line diagram; (b) Vector diagram.

fine tolerances and thus the necessity for two very stable and symmetrical amplifiers is obviated, if the relationship of the transformational accuracy of the properties of analogue amplifiers is smaller than with the previously described drive types 1 and 2. Even then, where very small signals have to be amplified the costs are smaller, since only low powers are necessary at the amplifier output, so that integrated circuits can be used.

Of the numerous switching systems for resolver digital converters mentioned in the literature, some which have practical applications in machine tool measurement systems may be picked out.

### 3.12.2.4  *The phase controlled method (drive type 2)*

In phase control processes [141–144;146], both stator windings use a rotary field that induces an alternating current in the rotor winding, the phase of which depends upon the position of the rotor:

$$u_3 = k_3 \hat{u} \sin(\omega t + \varphi) \tag{88}$$

The phase shift $\varphi$ between the input signal $u_1$ and the output signal $u_3$

is directly proportional to the angle of rotation $a$ of the rotor. For digital measurement of angle $a$, the resolver is adapted to a digital system, which the excitation voltage produces and the output signal evaluates. The excitation is produced via a filter, which removes the upper harmonics from the square-wave signal generated by the electronics, leaving the fundamental wave. By means of a null-seeking detector, the sinusoidal output signal of the resolver is transformed into a square-wave signal, the phase of which is related to a digital reference signal which contains the angle information, Figure 3.136. For the time interval between the positive flanks of these two signals a logic circuit opens a gate for a high-frequency timing signal which controls a counter. The output of the counter, with a fixed correspondence to the timing frequency, the frequency of the resolver excitation and the reference signal, is equal to the digital value of angle $a$. In order to obtain a periodic evaluation of phase shift, further control timing is necessary to transform each counter output in a buffer store and for subsequent zero-setting of the counter. In terms of control technology, this can be described as a type of digitization by sampling process with a holding term. All necessary timing signals are obtained from a single quartz-stabilized oscillator by frequency division. With a resolver frequency of 2.5 kHz and division of a period into 1000 steps, a timing frequency of 2.5 MHz is necessary.

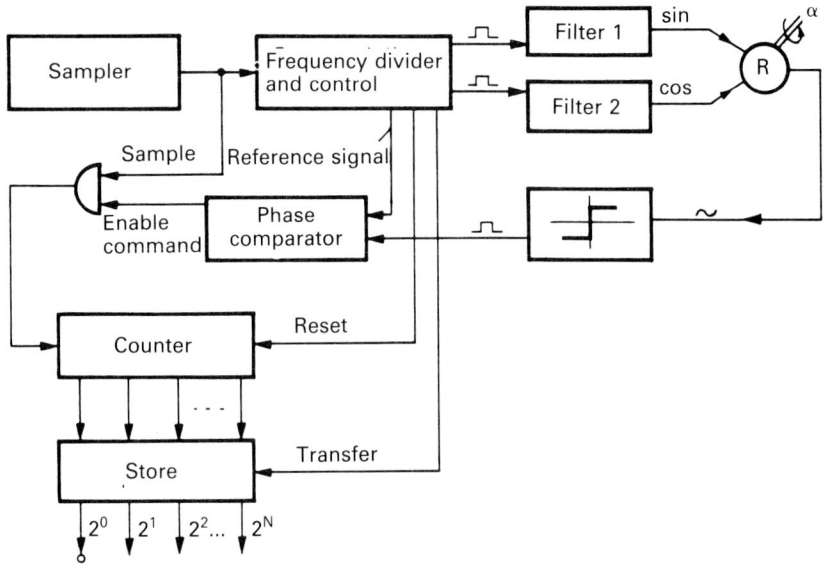

Figure 3.136   Position sensing using the phase control method.

Although this is an incremental measurement process, since the measured value is obtained via enumeration, it can also, because of its high redundancy, be regarded as an absolute method. The redundancy exists because of the periodic repetition of the measurement process. The problem of exact symmetry of the two filters for excitation voltages is capable of elegant solution, if the resolver is fed with digital signals the spectrum of which contains only higher-order harmonics. These harmonics are in fact carried through to the output signal and are there removed by a single filter which is switched before the null-seeking detector.

Figure 3.137 shows the principal processes for obtaining an excitation voltage for the resolver as a consequence of a square-wave signal. The initiation is a square-wave vibration with pulse-width repetition rate 1 : 1. Fourier analysis gives the following series:

$$u(t) = 0.5\hat{u} + \frac{2}{\pi}\hat{u}\left(\cos\omega t - \frac{1}{3}\cos3\omega t + \frac{1}{5}\cos5\omega t - \dots\right) \qquad (89)$$

The constant-voltage term $0.5\hat{u}$ is remotely maintained by galvanic division from the resolver and does not therefore need to be observed.

If the example of the third harmonic is taken, it is possible very simply to suppress the harmonics adjacent to the fundamental wave. This means that from the original signal with the condition 0 and 1 only partial pulses can be faded in or out.

Figure 3.137 shows a set of four pulses which have their centre at the place where the third harmonic reaches a maximum. Thus the spectrum of this set of pulses has its lowest frequency the third harmonic of the fundamental of the excitation voltage of the resolver. The amplitude of this voltage should be made so great in relation to the width of the four pulses that by superposition of the square-wave signals $a$ and $b$ the third harmonic is compensated. For the amplitude of the third harmonic of $u_1$ we have:

$$a_3 = \frac{2}{3\pi}\hat{u} \qquad (90)$$

For the harmonic of the set of pulses in Figure 3.129 we have:

$$a_3 = \frac{2}{T}\int_0^T f(t)\cos3\omega t\,dt$$

If the integration boundaries for the first pulse are defined as $30° + a$ (lower limit) and $90° - a$ (upper limit) and integrate as in the above expression, then we obtain:

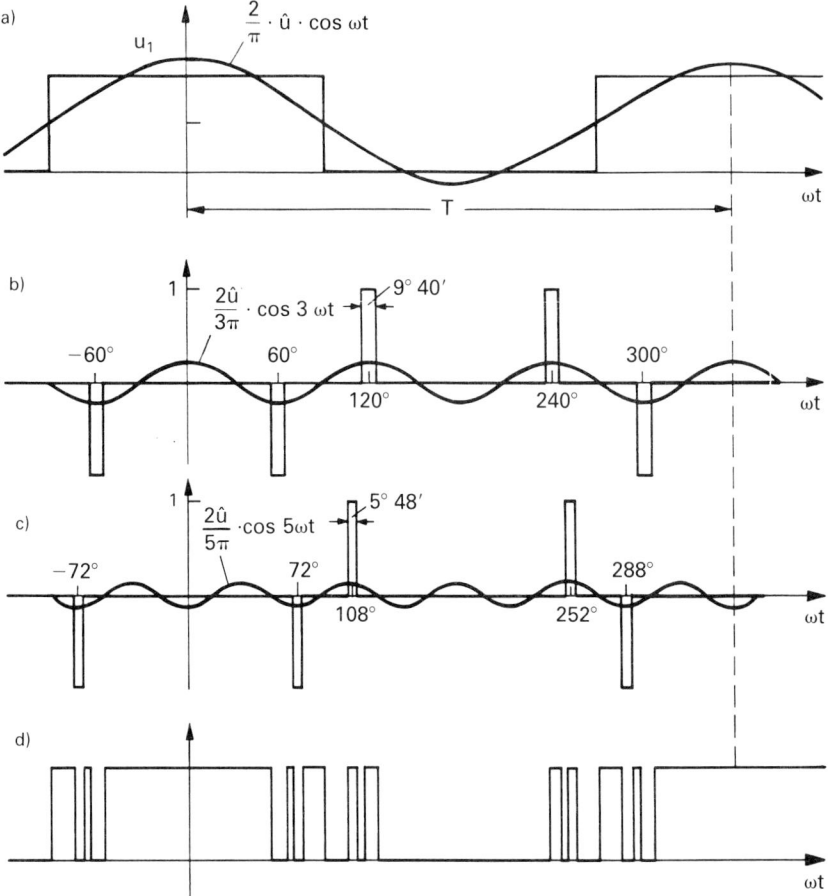

Figure 3.137   Digital exciting voltages. (a) Square-wave signal with fundamental wave. (b, c) Correction signal for compensating the 3rd or 5th harmonic. (d) Resulting square-wave signal. $u_1$, Fundamental wave of square-wave signal; $\hat{u}$, amplitude of square-wave signal; $T$, period of fundamental wave.

$$a_3 = \frac{8}{3\pi}\hat{u}\cos 3a \qquad (91)$$

Making $a_3$ equal to $a'_3$, a value of $a = 25°\ 10'$ is obtained, and thus a pulse width of $9°\ 40'$.

In this way it is possible to obtain a complete expression for the third harmonic.

If the process is expanded to include the fifth harmonic, where each period only adds two positive and two negative pulses, then we obtain:

$$a_s = \frac{8}{3\pi}\,\hat{u}\cos5\beta \qquad (92)$$

The factor $\beta$ also serves to define the integration limits. For the first negative pulse these are $(54° + \beta)$ and $(90° - \beta)$, where the centre of the pulses is $72°$. With $\beta = 6'$, the pulse width is $5° 48'$. In this way the fifth harmonic can be completely defined.

A newer method for processing sinusoidal vibrations without higher harmonics, where there is a phase difference of exactly $90°$, involves reading out from a set of fixed values held in read-only memory a set of discrete amplitude values in digital form, and converting them to digital voltages via a digital-to-analogue converter, Figure 3.138. Such devices are now cheaply available.

### 3.12.2.5   The amplitude controlled method (drive type 1)

In contrast to phase control processes, in which the two excitation voltages use a rotating field, amplitude control processes have both excitation voltages in phase, but with differing amplitudes.

The angular information is contained in the ratio of the two voltages. Since the output signal is zero only at the angular position defined by the amplitude ratio, this process is essentially a zero-seeking system, in which either the amplitude ratio is changed to the point where the output signal becomes zero (electrical zero-seeking system for angle measurement) or the rotor is mechanically moved to zero (mechanical zero-seeking system for position measurement). The analogue-to-digital conversion is carried out via a regenerative analogue-to-digital converter, Figure 3.139.

As a comparison term the resolver is taken with an angular position $a$. Of the many switching series tried in practice, the two basic digital-to-analogue converters should be compared for use as excitation voltage generators. The best-known processes for generating amplitude modulated alternating current use tapped transformers, which divide a constant primary voltage into steps corresponding to the sine or cosine of the defined angle. Switching systems for combinations of various taps are also available. Transformers and switches are, at the same time, function generators and multipliers for the input signal $u_0$. Since the resolver itself is a special form of

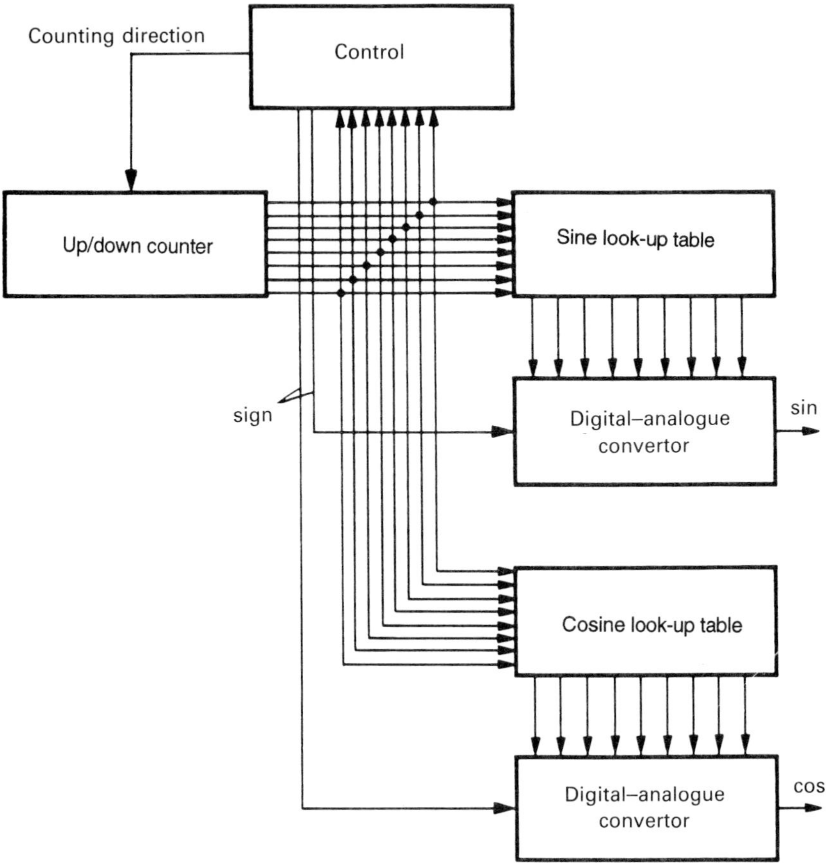

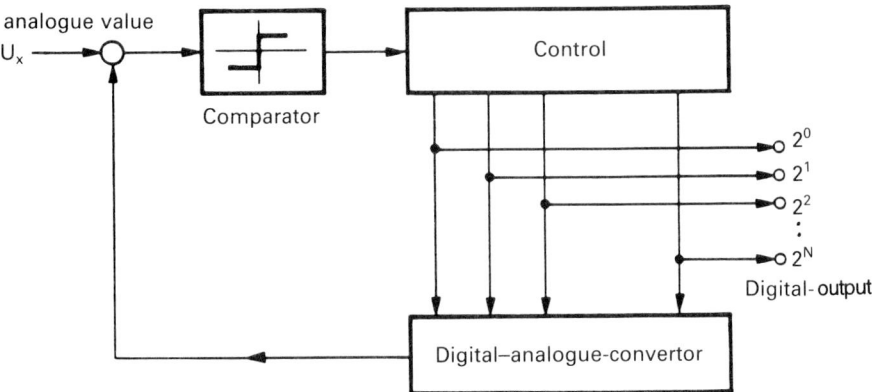

Counting direction

Control

Up/down counter

Sine look-up table

sign

Digital–analogue
convertor

sin

Cosine look-up table

Digital–analogue
convertor

cos

Figure 3.138   Sine–cosine generator using read-only memory (ROM) with function tables.

analogue value

$U_x$

Comparator

Control

$2^0$

$2^1$

$2^2$

$2^N$

Digital-output

Digital–analogue-convertor

Figure 3.139   Analogue-to-digital converter using the control loop principle with a digital-to-analogue converter.

Table 3.9 Transmission ratios and switching for 10 angles increasing by 36° steps

| $\varphi$ | Transmission ratios | | Switch |
|---|---|---|---|
| | $T_1$ | $T_2$ | |
| 0° | sin 0° | cos 0° | $S_3;S_6$ |
| 36° | sin 36° | cos 36° | $S_2;S_7$ |
| 72° | sin 72° | cos 72° | $S_1;S_8$ |
| 108° | sin 108° = sin 72° | cos 108° = −cos 72° | $S_1;S_9$ |
| 144° | sin 144° = sin 36° | cos 144° = −cos 36° | $S_2;S_{10}$ |
| 180° | sin 180° = sin 0° | cos 180° = −cos 0° | $S_3;S_{11}$ |
| 216° | sin 216° = −sin 36° | cos 216° = −cos 36° | $S_4;S_{10}$ |
| 252° | sin 252° = −sin 72° | cos 252° = −cos 72° | $S_5;S_9$ |
| 288° | sin 288° = −sin 72° | cos 288° = −cos 72° | $S_5;S_8$ |
| 324° | sin 324° = −sin 36° | cos 324° = −cos 36° | $S_4;S_7$ |

transformer, but with a mechanically variable transmission ratio, digital-to-analogue converter using a tapped transformer can also be interpreted as a form of resolver. Thus the transmission ratio is controllable by electrical signals. If the rotation of the resolver is divided into 1000 steps and three series-switched transformers are used, then the first has a sensitivity of 36°, the second of 3.6° and the third of 0.36°. Figure 3.140 and Table 3.9 show an example where eleven taps and eleven switches are used to give sine and cosine values for 36° steps overall [119]. To make up a system which splits the 360° into 0.36° steps, a three-stage arrangement of tapped transformers is necessary. This high cost of transformers and switches can be reduced if a transformer network is used based on the addition theorem for three angles $a$, $\beta$ and $\gamma$, where $a$ is the 36° partial angle, $\beta$ the 3.6° partial angle and $\gamma$ the 0.36° partial angle [122;123], Figure 3.133. The total angle is therefore:

$$\varphi = K_1 a + K_2 \beta + K_3 \gamma \qquad (93)$$

where $k_i = 0, 1, 2 \ldots 9; i = 1, 2, 3$.

Since, from the tapped transformer the sine or cosine function of the angle is subtracted rather than the angle itself, it is necessary to use, in combination, the addition theorem:

$$\begin{aligned} \sin\varphi = \sin(a + \beta + \gamma) &= \sin a \cos\beta \cos\gamma - \sin a \sin\beta \sin\gamma \\ &+ \cos a \sin\beta \cos\gamma + \cos a \cos\beta \sin\gamma \end{aligned} \qquad (94)$$

$$\begin{aligned} \cos\varphi = \cos(a + \beta + \gamma) &= \cos a \cos\beta \cos\gamma - \cos a \sin\beta \sin\gamma \\ &- \sin a \sin\beta \cos\gamma - \sin a \cos\beta \sin\gamma \end{aligned} \qquad (95)$$

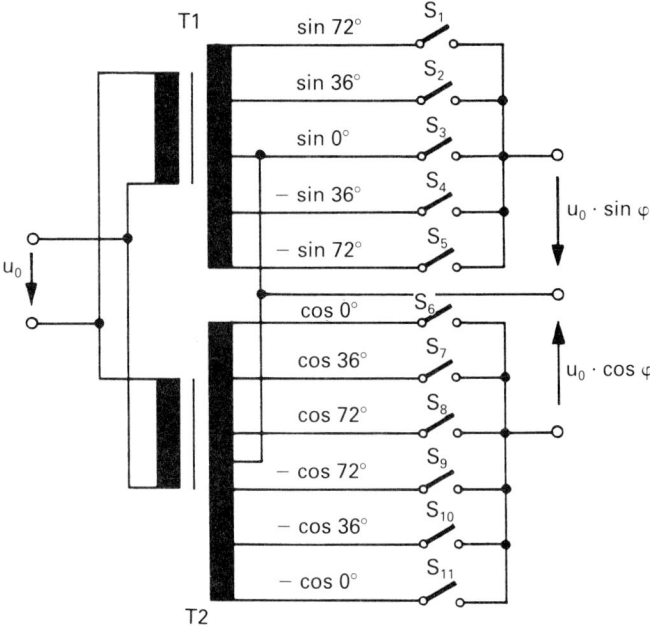

Figure 3.140   Transformer circuit for generating the functions $u_0$ sin and $u_0$ cos $\varphi$. S, Switch; $u_0$, voltage; $T_1$, $T_2$, transformers.

Substituting and factorizing, the following equations are obtained:

$$\sin\varphi = \cos\beta\cos\gamma[\sin\alpha + \cos\alpha\tan\beta + \tan\gamma(\cos\alpha - \sin\alpha\tan\beta)] \quad (96)$$

$$\cos\varphi = \cos\beta\cos\gamma[\cos\alpha - \sin\alpha\tan\lambda - \tan\gamma(\sin\alpha + \cos\alpha\tan\beta)] \quad (97)$$

The expressions in square brackets are obtained from switching (Figure 3.141) and give $T_2$, and $T_3$ as transmission ratios of the tangent values of the 3.6° steps, and $T_4$ and $T_5$ as transmission ratios of the tangent values of the 0.36° steps.

The product cos $\beta$ cos $\gamma$ can vary in value between 1 and cos(9 × 3.6°) × cos(9 × 36°) = 0.805. Since it occurs neither in the expression for cos $\varphi$ nor in that for sin $\varphi$, and is only dependent upon the ratio of these functions, it can be neglected.

Another form of expression for control voltages using tapped transformers is shown in Figure 3.142. Here, two transformers are used, of which the first has eight tappings, which give sine functions for angles 0–45° in 5.625° steps. The second transformer gives a linear interpolation between these values. For cosine functions in the range 0–45°, two further transformers are needed. Figure 3.143 shows the shape of the function for sine functions of 0–45° and the error curve, which is limited by linear interpolation [124;125]. Very

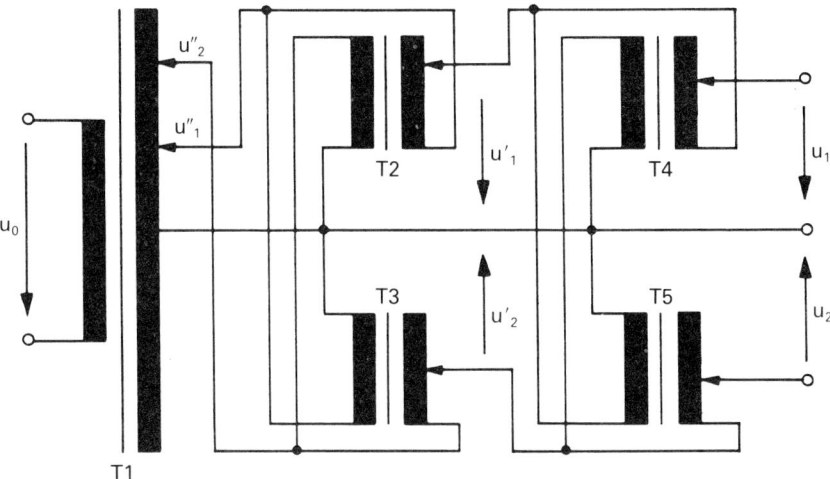

Figure 3.141 Cascaded transformers for generating exciting voltages by means of the addition theorem of angle functions. $u'_1$, Intermediate voltages; $u''_1$, intermediate voltages.

high accuracies are available via this method. Extension of the range between 45° requires a logic circuit which controls a system of switches in relation to each octant. The principle is shown in Figure 3.144. First, a control system is necessary that operates the switches on transformer taps so that the voltage curve $u_s$ and $u_c$ are produced (Figure 3.144). If, in addition, two converters are provided to change the polarity of the input voltage for both transformer combinations and two converters for alternatively coupling resolver windings 1 and 2 on both outputs, then four converters or eight switching signals (to operate relays) are necessary.

Control voltage generation using tapped transformers leads to very accurate results, since transformers are manufactured to higher accuracies. In these cases angular cores of low-loss iron are used. The accuracy of the transmission ratio of the transformer therefore is independent of temperature, ageing, etc. For switching, semiconductor switches with low 'on' resistance are used. The only disadvantage of transformers in this application is the high cost of precision transformers.

### 3.12.2.6 Digital sine–cosine generators (DSCG)

The digital sine–cosine generator process [121;126] is used for generating frequency modulated square-wave excitation voltages.

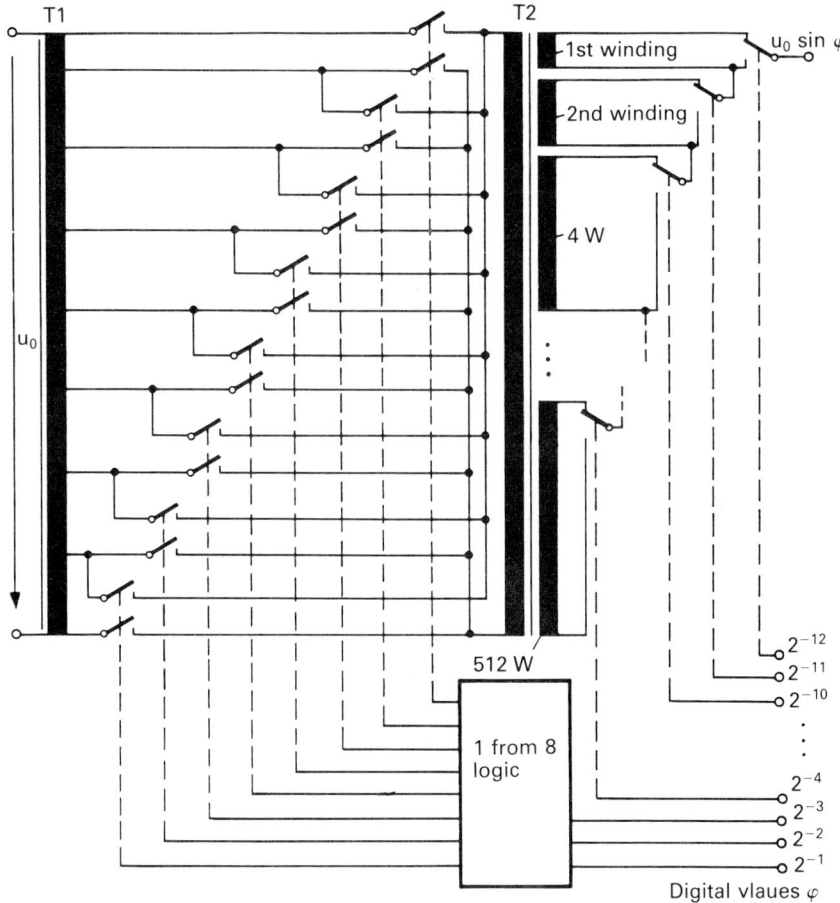

Figure 3.142   Formation of function $u_0 \sin \varphi$ between $\varphi = 0°$ and $\varphi = 45°$. $u_0$, Input voltage.

This frequency modulation is controlled by four pulse steps, as shown in Figure 3.145a.

For this set of pulses the following conditions apply:

$$T_2 \text{ lags } T_1 \text{ by } 1/4 \text{ period;}$$
$$T_4 \text{ leads } T_3 \text{ by } 1/4 \text{ period.}$$

Between these groups of signals, designated $T_1$ and $T_2$, there is a phase shift of $2\varphi - 1/4$ period.

If for the combination of $T_1$ and $T_3$ a further square-wave signal is used (Figure 3.145b), then examining its harmonic, a Fourier series is obtained

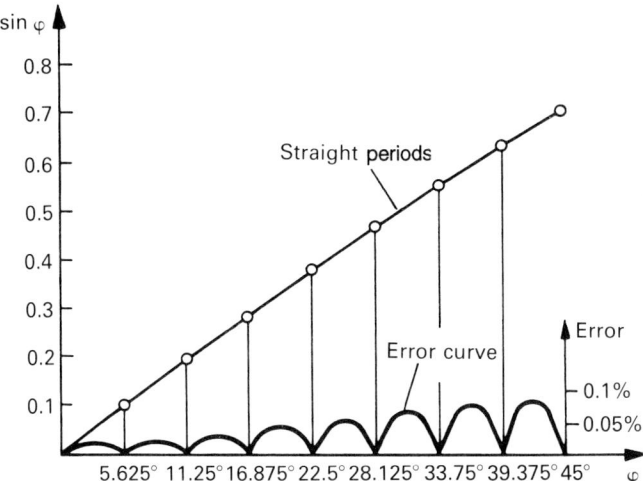

Figure 3.143   Error which arises when the sine function is replaced by a polygon train. a, Straight-line sections; b, error curve.

$$u_1 = \frac{2a}{\pi} \sin\varphi \cos\omega t$$

when the square-wave pulse has a height $a$ and width $2\varphi$, as well as being symmetrical about the zero time-point under consideration.

The function

$$u_2 = \frac{2a}{\pi} \cos\varphi \cos\omega t$$

is the harmonic of a square-wave function, Figure 3.145, of pulse width $(180° - 2\varphi)$ and a phase shift of about $T/2$ relative to the pulse series of Figure 3.137b. The second pulse series is obtained by combination of $T_2$ and $T_4$.

Functions $T_1$–$T_4$ and those derived from them can be defined by two counters $Z_1$ and $Z_2$, the antivalent outputs of which are each connected to a flip-flop, Figure 3.146.

The phase shift $2\varphi$ is controlled in relation to the output signal of the resolver. Thus the DSCG system is a zero-seeking system in which the resolver output signal is made equal to zero. Since the resolver output signal shows error compensation, it is often described as an error signal. Its evaluation is obtained according to sign and magnitude. For this reason the harmonics arising from excitation by square-wave signals are first of all filtered out. Then the fundamental

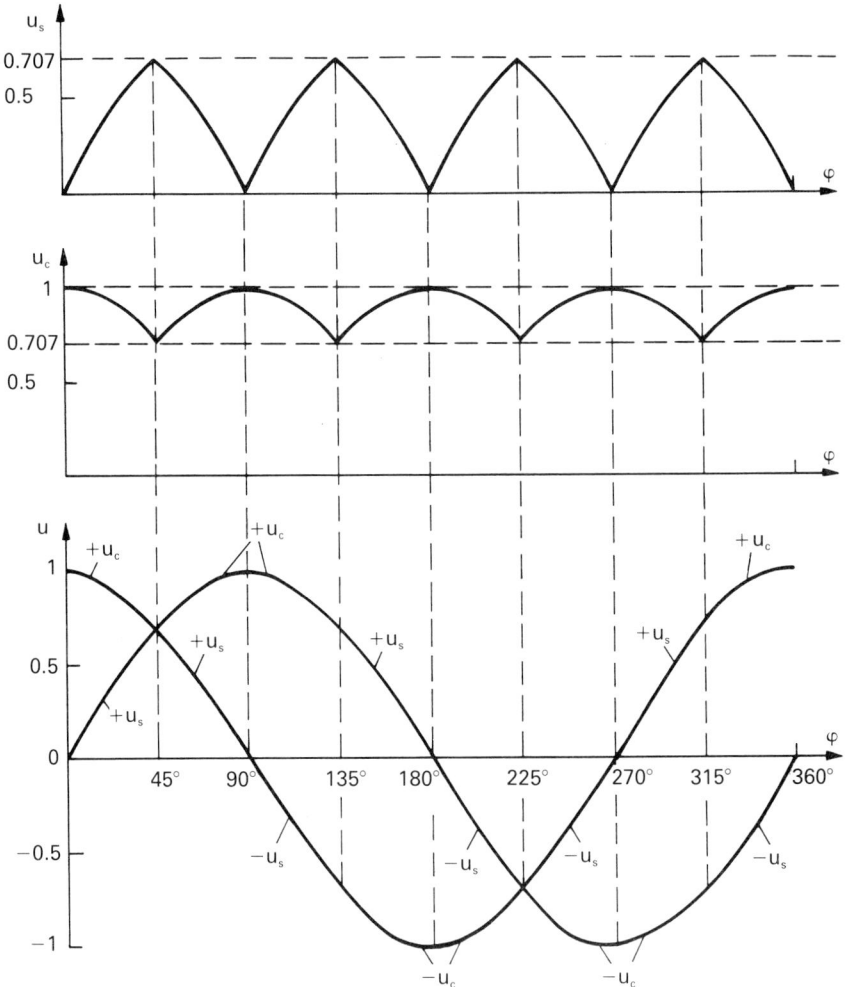

Figure 3.144   Synthesis of sine and cosine function from partial functions.
$u_S$, Sine voltage; $u_C$, cosine voltage.

signal is compensated by a phase-sensitive reference signal. Even harmonics of the working frequency or a 90° component are ineffective
after the compensation.

The sign differs depending on whether the output signal of counter
1 or 2 must be phase-shifted relative to the reference signal. The
phase shift is therefore affected, since in relation to the magnitude of
the error signal the input count for the particular counter is blocked
for a fixed time. The alternating control of both counters is necess-

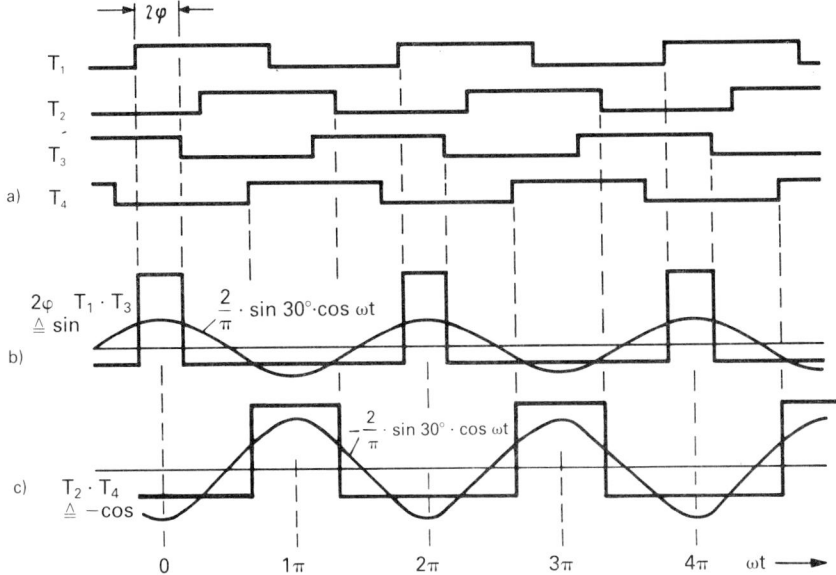

Figure 3.145   Obtaining the excitation voltage by means of pulse-width modulation. $T_1$–$T_4$, Control pulses.

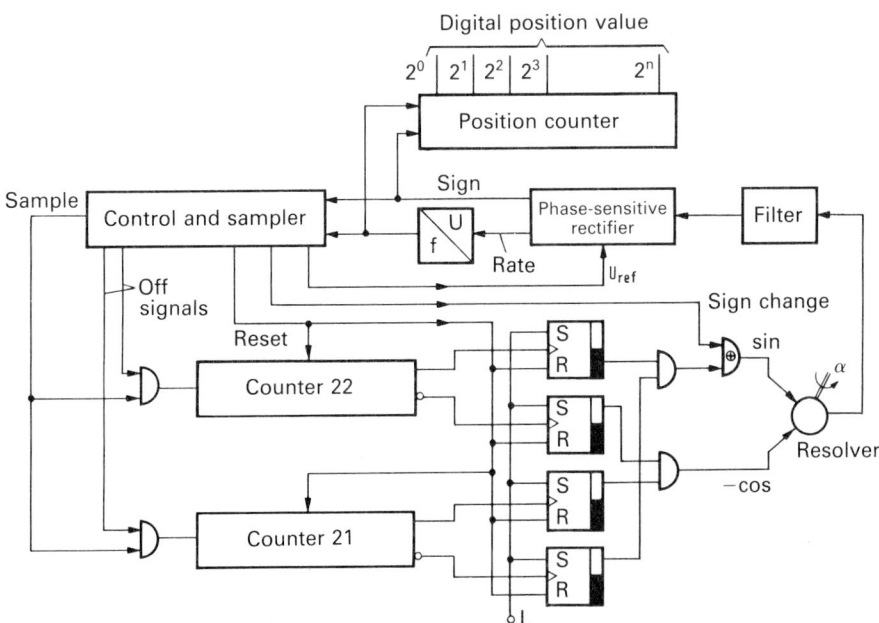

Figure 3.146   Basic principle of the digital sine–cosine generator.

ary, since by cutting off the count only negative phase shifts, or phase reduction, are permitted. As the number of cut-off pulses is a measure of the change in the magnitude of $\varphi$, it can simultaneously be used as a position indication. Impulse counting accurate in sign is therefore necessary. The disadvantage of this method of obtaining digital position information is a completely incremental signal evaluation.

### 3.12.2.7 Rotor winding feed—signal take-off from stator windings (drive type 3)

For a given angle $a$ between rotor and stator, there is for a feed to the rotor of a sinusoidal voltage $u_1 = \hat{u}_1 \sin\hat{\omega} t$, a voltage in the stator windings

$$u_2 = k\hat{u}_1 \sin\omega t \cos a \tag{98}$$

$$u_3 k\hat{u}_1 \sin\omega t \sin a \tag{99}$$

The ratio $u_3/u_2$ then contains the angular information. There are many processes for determining $a$, of which the most important, the null-seeking process, can be divided into two groups, integration processes and scanning processes.

In the integration process, output signals $u_2$ and $u_3$ are phase-sensitively rectified, multiplied by an auxiliary function which is the sine or cosine of $\varphi$, and then added and integrated. The integration term has the advantage that short-period noise pulses and harmonics are virtually eliminated. The disadvantage is the long integration time of at least half a period of the resolver voltage, restricted by the multiplex drive of the converter. In the scanning process within the same period and amplitude of a sine and cosine function are each sampled and stored until the end of the evaluation. Since the evaluation of the amplitude ratio can occur in parts of a single period, the scanning process is considerably faster than the integration process. The difference is greater the smaller the basic frequency (400 Hz technology). The scanning process is also distinguished from the integration process mainly in that neither a single filter nor an integration term is needed. For high accuracies (i.e. errors less than $0.36°$), with distortions due to noise or harmonics, and at high resolver speed, the integration process is to be preferred. In general, where very rapid changes in the measured value occur or where several systems are multiplexed to a converter, and at the same time

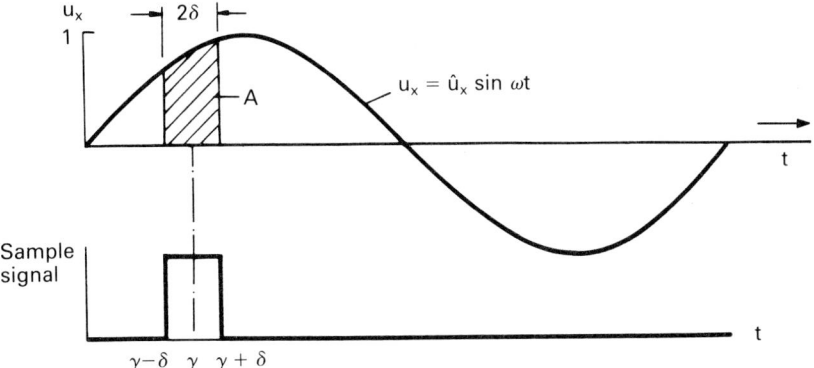

Figure 3.147 Scanning a sine function with a scanning time of 2. A, Surface; $\gamma$, scanning instant; $\delta$, half scanning time.

the resolution of the system does not exceed 10 bits, the scanner converter is preferable [127–130]. It can easily be shown that the effect of finite scanning period on the accuracy compared with the error sources experienced is small. If the value of a sine function for angle $\gamma$ is sampled, Figure 3.147, amounting to a width of sample impulse $2\delta$, then for surface $A$:

$$A = \int_{\gamma-\sigma}^{\gamma+\delta} u_x \mathrm{d}\omega t = 2\hat{u}_x \sin\gamma \sin\delta \qquad (100)$$

So long as $(\sin \delta)/\delta \approx 1$, a finite width of sample pulse leads to a negligibly small error. So, for example, the error is less than 1° per cent if $\delta \leqslant 4.5°$. Both groups of null-seeking processes, the integration as well as the scanning process, makes use of an auxiliary factor $\varphi$ to obtain a digital value for the angle.

We begin with the equation:

$$u_2 \sin\varphi - u_3 \cos\varphi = 0 \qquad (101)$$

Substituting equations (98) and (99) into equation (101) we further obtain:

$$k\hat{u}_1 \sin\omega t \cos a \sin\varphi - k\hat{u}_1 \sin\omega t \sin a \cos\varphi = 0 \qquad (102)$$

or

$$\sin(a - \varphi) = 0 \qquad (103)$$

The automatic balancing of this expression, which produces the auxiliary factor $\varphi$ an electrical angle which generates the mechanical angle, requires the following functions (Figure 3.148):

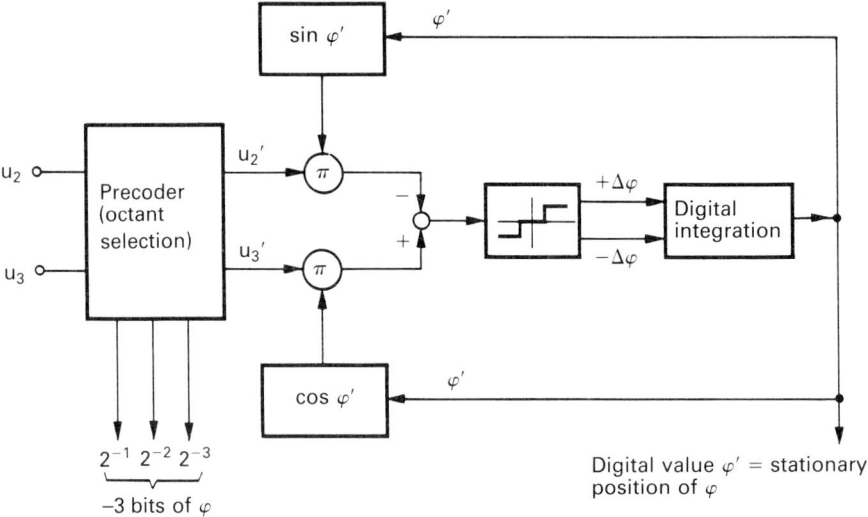

Figure 3.148    Functional modules of follow-up control methods.

- switching to select quadrant or octant,
- function modules for sine, cosine, or tangent,
- multiplexers,
- threshold switch,
- phase-sensitive rectifier,
- analogue subtractor,
- digital integrator or step-converter for successive approximations, apart from sampler and control apparatus.

At first sight the cost of this conversion method is considerable. However, it is possible to combine several function groups within one switching operation, as will be shown in the two following examples.

### 3.12.2.8    Converter with resolver bridge and octant encoder

The switching shown for the resolver bridge consists of two voltage dividers with electrically controllable partial ratios. Each of the two resolver signals is assigned to a voltage divider. The function of this voltage divider can be clarified by the simple example of an analogue sine potentiometer, in which the output signal is proportional to the sine of the angle of rotation. Since this is also proportional to the input voltage, the output signal represents the product of the input voltage and the sine of the angular position of the slider. Digital control of the partial ratios is necessary with a resolver bridge, for which reason the voltage distributor is provided with switchable taps. The steps are so arranged that by control of the digital value of $\varphi$ the partial ratio corresponds to the sine of $\varphi$ (Figure 3.149). Thus the function generator for the sine function and the multiplier are combined in a relatively simple manner. For the principle it is unimportant whether the voltage divider is resistive, inductive or capacitive. Inductive voltage dividers are in widespread use as tapped transformers (see Section 3.12.2.5 and Figures 3.135–3.137). The two voltage dividers of the resolver bridge are arranged in accordance with equations (98) and (99) so that one follows the sine and the other the cosine function. Subtraction of the output signals $u_2$ and $u_3$ is simply carried out by means of an operational amplifier. If equation (101) is not fulfilled, then the differential voltage operates a threshold switch, which alternatively allows one of two channels for the sign-oriented control of an integrator or a step-converter. The digital output signal $\varphi$ of the integrator is led to the resolver bridge. In the balanced condition the value of $\varphi$ is then equal to the value of the mechanical angle $a$. It is also possible to add an octant encoder to a resolver bridge, but it is not absolutely necessary [125]. If used, it limits the working range of the resolver bridge to an octant, or 45°, giving a notable reduction in cost. The octant encoder gives coded information over each octant, in which the three bits of the angle to be measured with the highest value are known. The differences over each octant in terms of the polarity and the amplitude ratios of $u_2$ and $u_3$ are shown in Table 3.10.

Table 3.10   Criteria for the selection of octants

| Octant | sign $u_2$ | sign $u_3$ | |
|:---:|:---:|:---:|:---:|
| 1 | +1 | +1 | $\lvert \hat{u}_2 \rvert > \lvert \hat{u}_3 \rvert$ |
| 2 | +1 | +1 | $\lvert \hat{u}_2 \rvert < \lvert \hat{u}_3 \rvert$ |
| 3 | −1 | +1 | $\lvert \hat{u}_2 \rvert < \lvert \hat{u}_3 \rvert$ |
| 4 | −1 | +1 | $\lvert \hat{u}_2 \rvert > \lvert \hat{u}_3 \rvert$ |
| 5 | −1 | −1 | $\lvert \hat{u}_2 \rvert > \lvert \hat{u}_3 \rvert$ |
| 6 | −1 | −1 | $\lvert \hat{u}_2 \rvert < \lvert \hat{u}_3 \rvert$ |
| 7 | +1 | −1 | $\lvert \hat{u}_2 \rvert < \lvert \hat{u}_3 \rvert$ |
| 8 | +1 | −1 | $\lvert \hat{u}_2 \rvert > \lvert \hat{u}_3 \rvert$ |

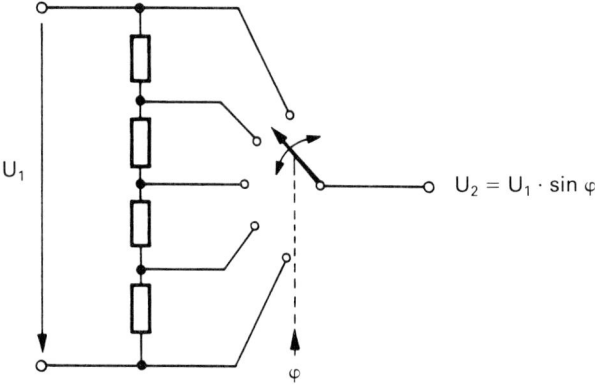

Figure 3.149   Voltage divider as multiplier and function generator.

The evaluation of Table 3.10 follows the four signals which must be obtained from the analogue input signal. They are described by the following equations:

$A$:   $\hat{u}_2 \geqslant 0$, corresponds to signal $u_2$     (104)
$B$:   $\hat{u}_3 \geqslant 0$, corresponds to signal $u_3$     (105)
$C$:   $\hat{u}_2 - \hat{u}_3 \geqslant 0$     (106)
$D$:   $\hat{u}_2 + \hat{u}_3 \geqslant 0$, gives inequality $|\hat{u}_2| <> |\hat{u}_3|$     (107)

As Figure 3.150 shows, the alternating voltages $u_2$ and $u_3$ are phase-sensitively rectified by means of a reference voltage $u_R$ and then smoothed via a low-pass filter. Four comparators are then used for digital signals $A$, $B$, $C$ and $D$. A converter amplifier is necessary for the production of $-u_3$. The control signals for the alternating converter are obtained from the following expressions, Figure 3.151:

$$T_1 = A\bar{C} + A\bar{D} = A(\bar{C} + \bar{D}) \tag{108}$$

$$T_2 = BC + B\bar{D} = B(C + \bar{D}) \tag{109}$$

$$T_2 = C\bar{A} + \bar{A}D \tag{110}$$

$$T_4 = D\bar{B} + \bar{C}\bar{B} \tag{111}$$

$$T_5 = A\bar{C} + D\bar{A} \tag{112}$$

$$T_6 = CD \tag{113}$$

$$T_7 = C\bar{D} \tag{114}$$

$$T_8 = \bar{C}\bar{D} \tag{115}$$

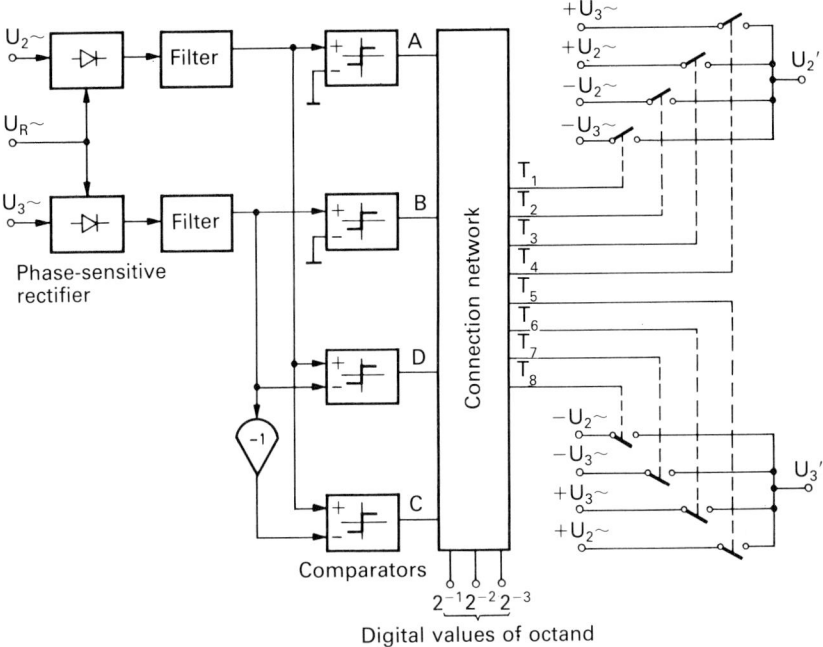

**Figure 3.150** Octant encoder. *A*, *B*, *C*, *D*, Square-wave signals derived from resolver voltages; $T_1$–$T_8$, control pulses.

For coding the octant the following rules apply:

$$2^{-3} = A\bar{D} + A''D + \bar{B}\bar{C} + BC = (A \oplus D) + (B \equiv C)$$

$$2^{-2} = A\bar{B} + \bar{A}B = A \oplus B$$

$$2^{-1} = \bar{A}.$$

The octant encoder can be used either as a converter according to the integration principle or as according to the scanning principle. In the latter case the input stage is set with a phase-sensitive rectifier and low-pass filter through a controlled switch and signals *A*–*D* sampled consecutively in a 1 bit inter-sampler, i.e. a D-FF.

### 3.12.2.9 *Converter with control integrator (demodulation method)*

Multiplication of the resolver signal by the sine or cosine of an auxiliary function $\varphi$ is possible through integration over limits dependent

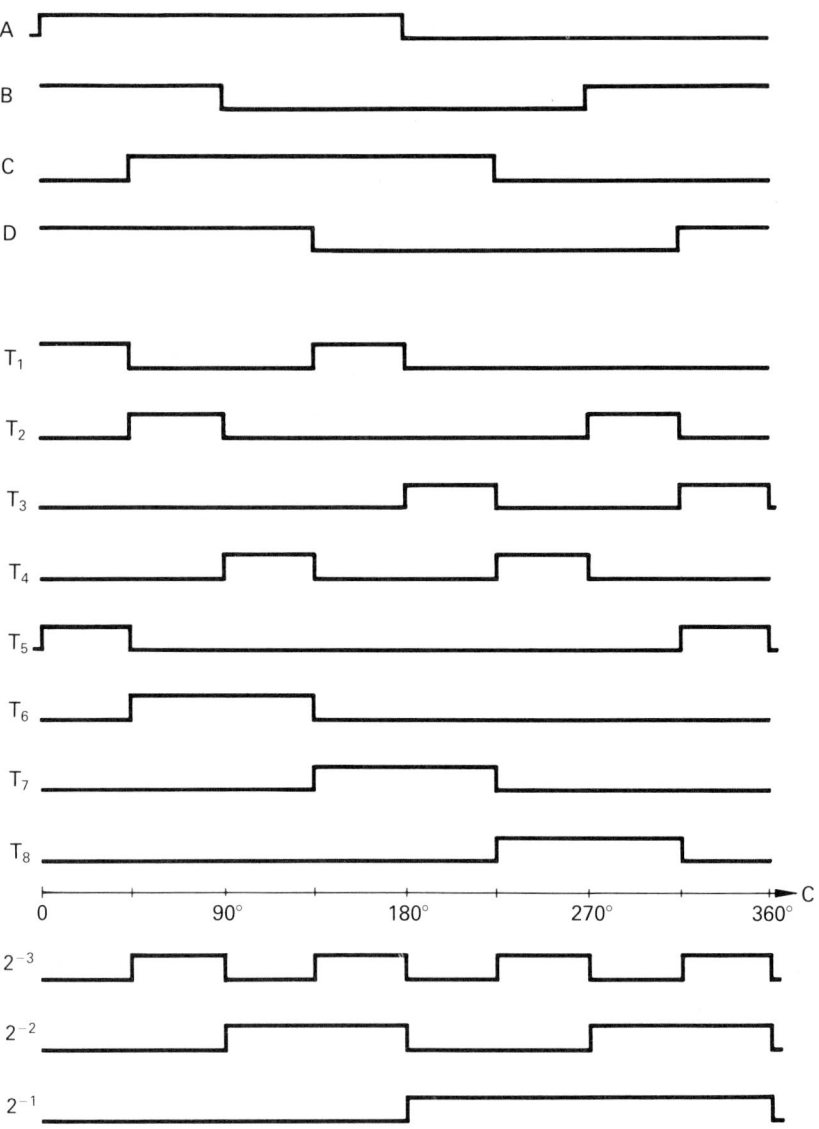

Figure 3.151    Clocking scheme for octant encoder. Phase angle.

on $\varphi$ [131;132]. Thus, in Figure 3.147, the hatched area under the sine function indicates that it is limited by the abscissa values $\gamma - \delta$ and $\gamma + \delta$. The expression for area $A$ is therefore:

$$A = 2\hat{u} \sin \gamma \sin \delta.$$

Equation (100) provides a solution for the balance condition of the resolver signal $u_2 \sin\gamma - u_3 \cos\gamma = 0$, equation (101), when $\gamma = \gamma$ and $\gamma = 90°$. The value $u_2$ must then be integrated over a range from $\varphi - 90°$ to $\varphi + 90°$. Since $-\cos\varphi = \sin(\varphi - 90°)$, $u_3$ must be integrated from $\varphi - 180°$ to $\varphi$, Figure 3.152a. In this example the mechanical angle is chosen as $a = 60°$, which corresponds to $\varphi$, since the balance condition is shown. The integration limits for $u_2$ are therefore $-30°$ and $+150°$ and for $u_3$ $-120°$ and $+60°$. From Figure 3.152 it can also be seen that variation of the integration process is possible, since according to the alternating direction of $u_2$ and $u_3$ with the signals $T_1$ and $T_2$ an integration with the fixed limits 0 and 180° occurs. The process is then shown as an inverter or phase converter process [132]. The measuring process then continues with the output signal of the inverter integrated over a half period. If it then gives a positive or negative constant voltage value, then the phase of cycles $T_1$ and $T_2$ respectively, alters the resolver signal so that in the next period, the balance condition is fulfilled. Thus, one of the output signals of the integrator is proportional to the number of pulses of a network, where correspondingly many pulses are input or output from the basic cycle of the counter. The output signal of a successive counter that receives an input impulse of, say, $1:1000$ is therefore first shortened or lengthened and thus undergoes an actual phase change. Figure 3.153 shows the principle of using this phase change. The step-down-ratio of the counter in the example is taken as $1:4$. The phase of signal $A_2$ corresponds to counter signal $T_2$. From signal $A_1$ the feed voltage for the resolver is used, in which according to an amplifier stage the harmonics are reduced through a filter, Figure 3.154. The reference signal is dependent on a resolver taken from a sinusoidal feed signal by means of a null-seeking detector. The required digital value of $\varphi$ exists in a differential counter in which the interval between the positive flanks of reference and control signal $T_2$ becomes what is counted by the counter. If the differential counter range is assumed to be 1000, then this means that the sensitivity to angle $\varphi$, which can be a maximum of one resolver rotation, is 1000 steps. After the end of a digitizing cycle the value of the differential counter is transferred to an intermediate buffer and the counter is then cleared. It does not matter whether the angle $a$ has changed, since the value of $\varphi$ is newly determined periodically with the digitization cycle. Because of this high degree of redundancy, the system can be regarded as an absolute measurement system. If the value of the differential counter is briefly falsified because of an error pulse, this error will be corrected in the next cycle.

Additionally, the simplest process for evaluating the amplitude

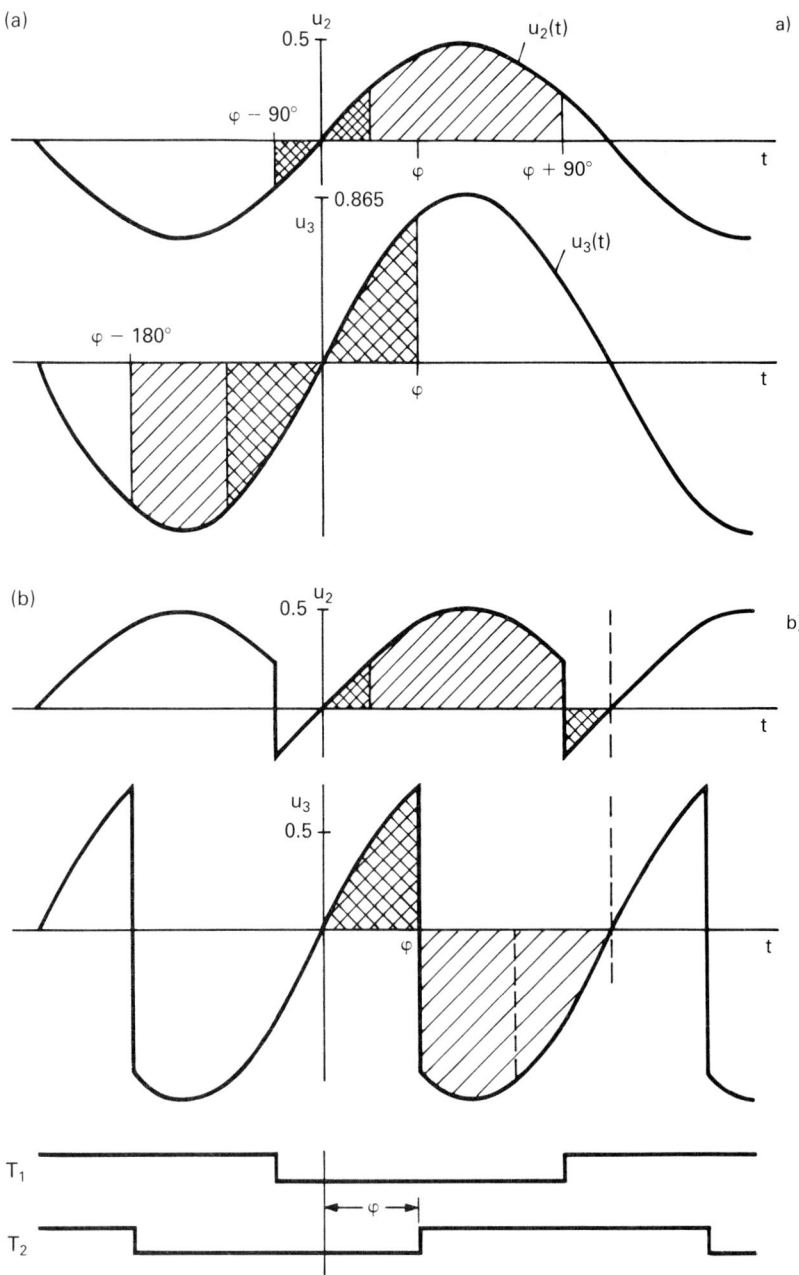

Figure 3.152   Demodulation method: signal diagrams. (a) Different integration levels for $u_2$ and $u_3$. (b) Common integration levels for $u_2$ and $u_3$. $T_1$, $T_2$, Control signals for inverter.

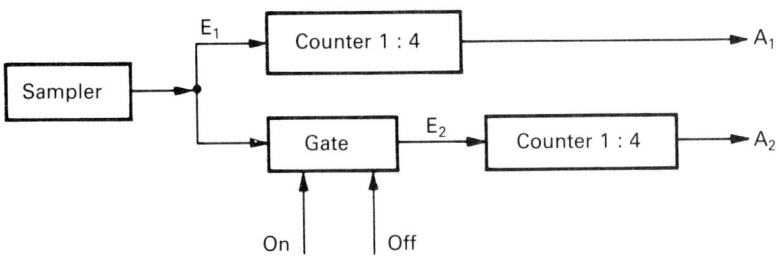

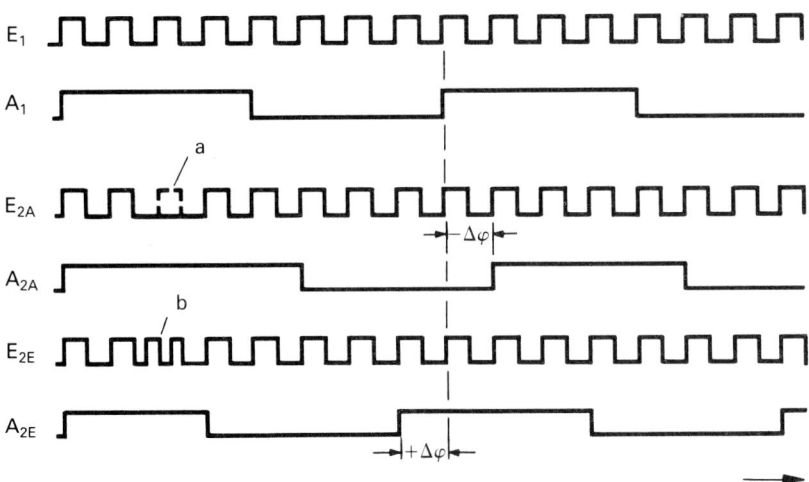

Figure 3.153　Digital phase-shifting by gating pulses in or out.
a　Pulse gated out
b　Pulse gated in
E　Gating in
A　Gating out

ratios of $u_2$ and $u_3$ should be mentioned, i.e. what is usually called the bridge method. Here the amplitude-modulated alternating voltages are formed into two phase-shifted voltages, with a phase difference equal to the required angle $\varphi$, Figure 3.155. It is assumed that the bridge arms are chosen such that $R = 1/\hat{\omega}C$.

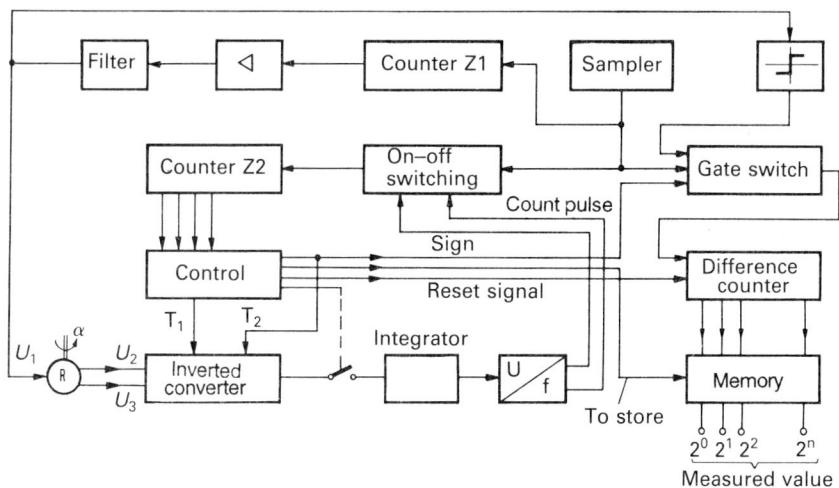

Figure 3.154   Inverter method.

$T_1$, $T_2$   Control signals for inverter.

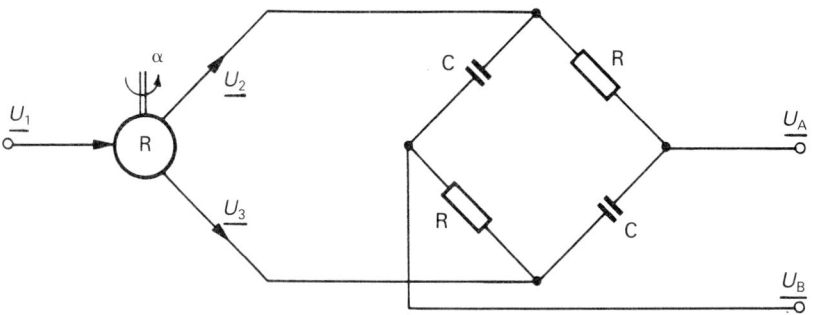

Figure 3.155   Converting amplitude-modulated resolver voltages into two phase-shifted voltages.

$U_1$ to $U_3$   Resolver signals

$U_A$, $U_B$   Phase shifted output voltages of the bridge

Then we have:

$$\underline{U}_A(t) = \frac{\underline{U}_2(t) - j\underline{U}_3(t)}{1+j} \qquad (116)$$

$$\underline{U}_3(t) = \underline{U}_2(t)\tan\alpha \qquad (117)$$

Equation (117) is substituted in equation (116) giving:

$$\underline{U}_A(t) = \frac{\underline{U}_2(t) - j\underline{U}_2(t)\tan\alpha}{1+j} \tag{118}$$

$$\underline{U}_A(t) = \underline{U}_2(t)\frac{1 - j\tan\alpha}{1+j}$$

The evaluation is given as a phase shift between $U_A$ and $U_1$ $\varphi_A = a - 45°$. In the same way a phase shift $\varphi_B = 45° - a$ is found between $U_B$ and $U_1$. The phase shift between $U_A$ and $U_B$ is thus $\varphi_A - \varphi_B = 2a - 90°$.

The complete switching set-up for evaluating the resolver signals is shown in Figure 3.156. Both comparators produce square-waved signals, the flanks of which defined the zero crossings of $U_A$ and $U_B$. With the post-switched monoflops, the positive flanks of this square-wave signal are used for downloading the counter or transferring the counter results into the buffer. As the output signal of the RS flip-flop shows one pulse-width that corresponds to the phase difference between $U_A$ and $U_B$, the counter output is precisely the desired value $2a - 90°$. The constant 90° is calibrated into the system, and the factor 2 eliminated via the counter. The great advantage of the bridge method is its simple principle. It counteracts the high requirements on the bridge components, the properties of which must be extremely constant over time and independent of temperature. The capacitors present the most serious problem in this respect. Further limitations on this process also occur in that interference peaks from the resolver signal can adversely affect the comparator signal. Since the output signal of the comparator is not synchronous with the counter, a rounding error of one bit can arise in addition to the unavoidable errors in quantization.

## 3.13 Inductosyn

The Inductosyn™ (the Inductosyn Corporation, New York, USA) measurement system is in effect a resolver extended into a plane [133;135;145]. Depending on its form, it can be used for angle measurement (circular Inductosyn) or for distance measurement (linear Inductosyn). In both cases it consists of two plane components made of non-magnetic material (glass, non-magnetic steel) which can move relative to one another, separated by an insulating intermediate layer of undulating conductors in the form of a printed circuit. The linear Inductosyn consists of a linear section and a rider. The linear section corresponds to the rotor, and the rider, which has two opposed windings on a quarter period, to the stator of

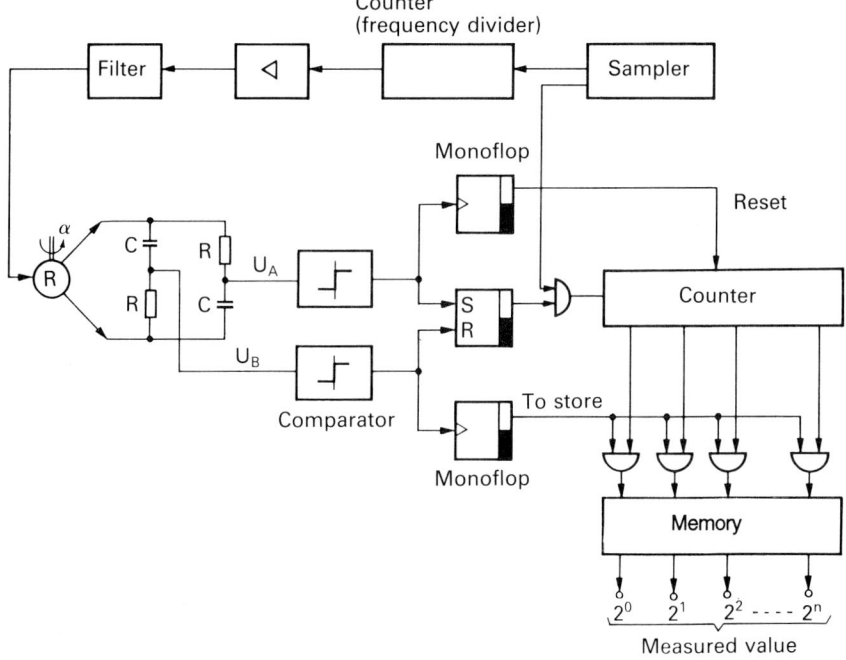

Figure 3.156   Digitizing resolver signals using an RC bridge.

a resolver, Figures 3.157 and 3.158. Thus the feed of the linear section to the rider voltage conforms to equations (73) and (74). Again, as with the resolver, the linear Inductosyn can be regarded as a special case of a transformer. Figure 3.159 shows how the magnetic field arising from the linear section in one winding of the rider induces a proportional voltage in the opposite layer. In the equivalent circuit diagram, Figure 3.160, only three quantities appear:

- the resistance of the linear section $R_1$,
- the mutual inductance $M$ proportional to the displacement $x$,
- the rider winding resistance $R_2$.

For a typical operating frequency of 10 kHz, therefore, the following values apply

$$R_1 = 3.3 \, \Omega, \ \omega M = 0.021 \, \Omega \text{ and } R_2 = 0.75 \, \Omega.$$

For the transmission factor we have $u_2/u_1 = k$. The magnitude $k$ has a value of 0.025 for an air-gap of 0.25 mm [136].

The linear Inductosyn is a well-developed direct measurement system for machine tools. The reason for this is the relatively small space requirement, the considerable insensitivity to fouling and the possi-

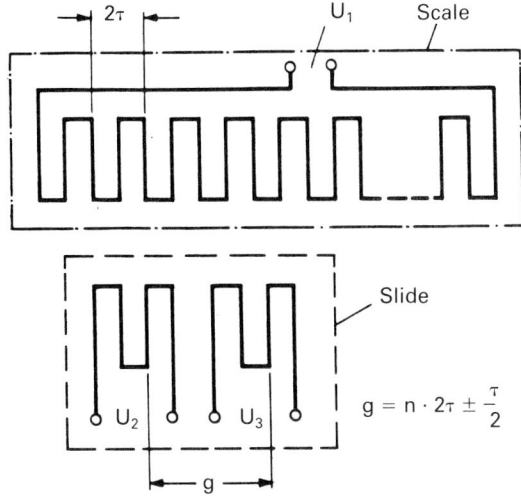

**Figure 3.157** Principle of the inductosyn measuring system.
$\tau$   Division
$g$   Distance between slide windings
$n = 1, 2, 3, \ldots$

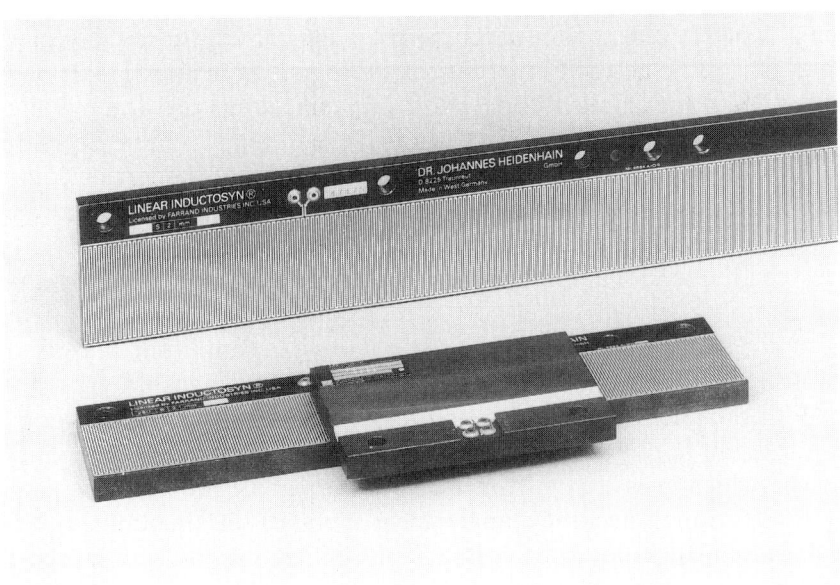

**Figure 3.158** Scale and slide of the linear inductosyn.
(Picture courtesy of Heidenhain).

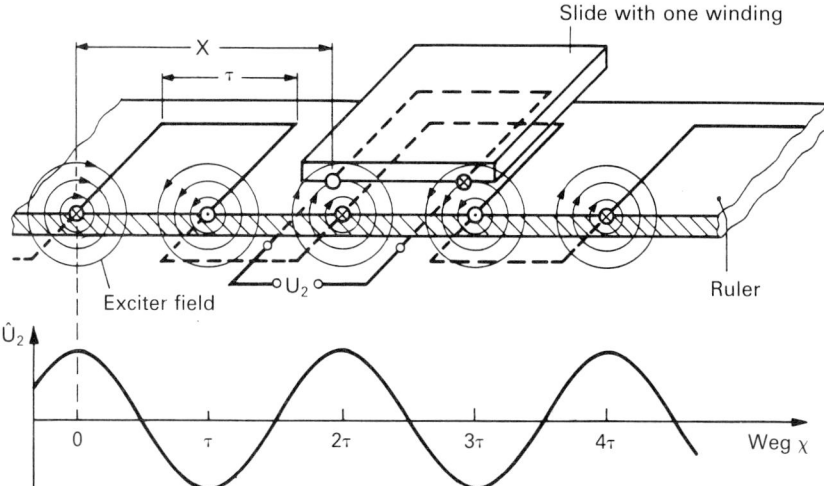

Figure 3.159  Magnetic coupling between scale and slide windings and amplitude response of slide voltages $\hat{u}_2(x)$.
$\tau$  Division

bility of measuring different lengths as required by inserting unit lengths of 250 mm each. The high accuracy is also adjustable to give three classes, which are, respectively, 5, 2.5 and 1 $\mu$m error [137].

With the mechanical connection of the scales, the windings are switched in series electrically. Although the measured value depends only on the relative movement between linear section and rider, it is also possible in machine tools to have the linear section fixed and the rider on the moving part of the tool, for space reasons. The air-gap between the sender and receiver windings must lie in the range 0.125–0.375 mm [138]. Variations in the region of ±0.5 mm do not affect accuracy, since they only produce variations in amplitude of $u_2$ and $u_3$ of ±10%, so that the amplitude ratio does not change.

Both the linear and circular Inductosyns come in various versions for different applications and accuracies. The original circular Inductosyn, which was at one time used as an analogue measurement system for high-accuracy angle measurements in theodolites and radar antennae, allows resolutions up to $10^{-6}$ rotations with an error of ±2 arc seconds. It consists of two flat discs of about 12 inches diameter, of which one is the stator while the other serves as a rotor by being fixed to the rotating machine component, Figure 3.161. Thus the rotor corresponds to the linear section of the linear Induc-

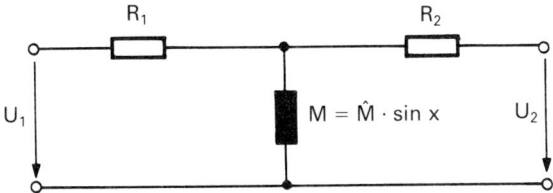

Figure 3.160    Equivalent circuit of the Inductosyn measuring system.
M    Peak value of mutual inductance

tosyn. Each winding forms a pole, so that it forms systems with 2, 144, 256, 360 and so on up to 2000 poles. The stator carries several separated groups of windings, which are switched so that two output signals arise, and this corresponds to the rider of the linear Inductosyn.

For these signals we have

$$u_{2R} = K_R \hat{u}_2 \sin\omega t \cos\frac{N}{2} a \qquad (119)$$

$$u_{3R} = K_R \hat{u}_3 \sin\omega t \sin\frac{N}{2} a \qquad (120)$$

where $K_R$ is the coupling factor for the circular Inductosyn, $N$ the number of poles and $a$ the angle of rotation.

The number of signal periods in the output voltage is thus equal to half the number of poles, so that a 2000 pole system produces 1000 signal periods. Compared with a resolver, therefore, the sensitivity is increased by a factor of 1000. Since the digitization has a resolution of 1/1000th of a period, the overall resolution is $10^6$ positions per revolution. The number of pairs of poles passing through during a measurement process is usually read off via a counter. The electrical excitation of the rotor occurs either via a special winding, as in a transformer, or through brushes.

The linear Inductosyn in its basic form has a scale of 250 mm in length and a pole separation of 1 mm (metric scale), and digitization is possible to a sensitivity of $2\,\mu$m. Since the Inductosyn is a cyclic absolute system, attempts have been made to produce an absolute measurement system by superimposition of three systems of different pole separations. These triple Inductosyns, as they are called, have pole separations of 1, 10 and 100 mm. Figure 3.162 shows the main layout of a coarse system, while Figure 3.157 gives the layout of a

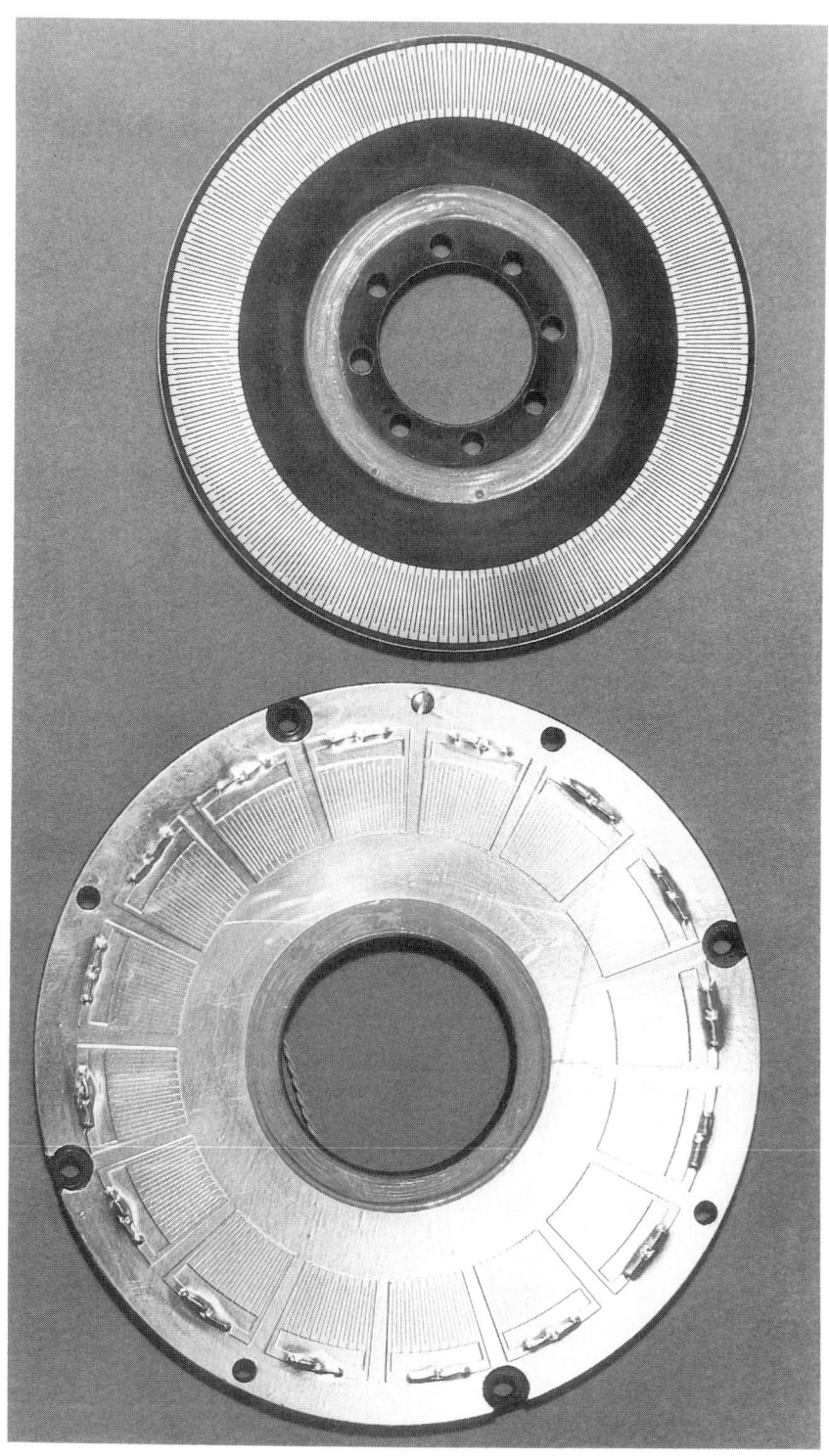

Figure 3.161    Rotor and stator of the round Inductosyn.
Picture courtesy Inductosyn Corp.

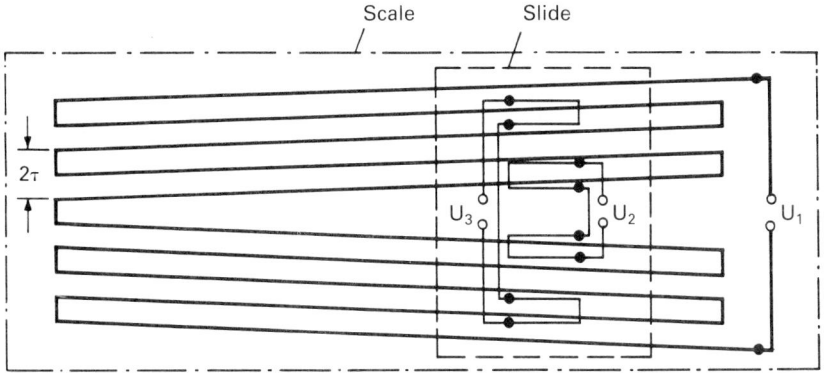

Figure 3.162   Arrangement of conductor tracks in the coarse system.

fine system. Although the main disadvantage of the Inductosyn is overcome here, namely that it is only absolute over a range of 2 mm, the triple Inductosyn is used only infrequently compared to the single system. The coarser tracks are usually replaced by a resolver, when not even an additional measurement system is saved, and the number of signal periods is read off via a counter. A problem that occurs with both circular and linear Inductosyns is the relatively narrow tolerance in construction and adjustment of the discs or scales. Examples of the requirements for construction of a linear Inductosyn have been quoted as [139]:

- maximum unevenness of surfaces of rider and linear section ±2.5 μm,
- maximum difference in height between individual scales measured at the air-gap ±50 μm,
- maximum variation of linear axis from moving axis ±50 μm,
- permissible air-gap between rider and scale 125–375 μm.

There are various processes for adjusting the scales in the longitudinal direction. The earliest processes involved gauge blocks of the length of the scale and clock gauges. Thus the position of each scale was always referred to that of the others. Because of this, it was impossible to avoid summing the adjustment errors. They can, however, be avoided by the use of an absolute comparison scale, e.g. a photoelectrically contacted scale, the length of which corresponds to the total length. Adjustment using a laser interferometer is, however, the simplest and most accurate process, which will be used almost exclusively in the future. Although adjustment of individual scales is time consuming and therefore costly, it allows compensation of individual scale errors within known boundaries, since the error sum curve can be referred to zero for each scale by deliberately setting them slightly off. Similar requirements apply for adjustment of circular Inductosyns [139].

## 3.14  Accupin

The Accupin™ (General Electric Co., New York, USA) [147] is an analogue sensor consisting of two quadrupoles, which has a transmission factor according to the following equation:

$$u_2 = \text{const} \times u_1 \sin(x + a) \tag{121}$$

Here, $u_1$ is the input alternating voltage, $x$ is the physical displacement between measuring head and scale and $a$ is a constant which at one quadrupole is zero and at the other a quarter period. Both quadrupoles are interconnected so that the system can have the properties of a resolver or an Inductosyn. It is therefore possible not only to use them under the same operational circumstances, but also to use the same digitization process. As can be seen in equation (121), it acts as a cyclic absolute measurement process. The use of amplitude-modulated output signals, however, works according to an entirely different principle, which at its most basic is comparable with that of the inductive sensor, Figures 3.12 and 3.15.

The distance to be measured is covered by an adjustable iron core, which changes the inductance of two coils. In an a.c.-fed bridge, the displacement of the iron core within a given range generates a diagonal voltage $U_D$ proportional to the magnitude and direction of the displacement. When the proportion range is exceeded, $U_D$ forms a first-approximation sine curve. In contrast to an inductive sensor, in which only the linear range is used, the Accupin can use the sinusoidal curve of the bridge current by use of specially designed coils and magnetic return path. The scale, which is made of sliding cylindrical pins carried on a bearer of non-magnetic steel, forms the magnetic return path. The pins, which in the metric version are 2 mm in diameter, are made of ferromagnetic nickel steel. The measurement head consists of four coils which are formed around laminated iron cores with an air-gap. The sheet stack is made up of non-magnetic intermediate layers (Figure 3.163) separated by a distance corresponding to the pin diameter, in order to obtain a sinusoidal change of the amplitude of the bridge current with displacement of the measurement head. For each of the two quadrupoles there is bridge switching, which involves two coils of the measurement head which are displaced relative to each other by half a pin spacing. Thus the impedance of one coil reaches a maximum when the other reaches a minimum, and the bridge has maximum sensitivity. Since the output signal of one bridge must follow the sine and the other cosine of the displacement distance, the coils of the first bridge are displaced one quarter period relative to the second. The usual application of the Accupin is a phase-controlled process. The feed voltage for bridge 1 is $u_1 = \hat{u}_0 \sin\omega t$ and for bridge 2 is $u_2 = \hat{u}_0 \cos\omega t$. The arrangement of the measuring head inductances in both bridges is selected so that the voltages follow the relationship $\hat{u}_D1 = K \cos x$ and $\hat{u}_D2 = K \sin x$, Figure 3.164. If both diagonal voltages are added then we obtain:

$$\hat{u}_3 = \sqrt{\hat{u}^2_{D1} + \hat{u}^2_{D2}} = K \tag{122}$$

representing a signal of constant amplitude which contains the distance information in its phase. One cycle of the phase modulation thus corresponds to a movement which is equal to the pin displacement. The feed to the two bridges is not, as is theoretically possible, from one fixed two-phase network, but from two differential resolvers, the rotors of which are rigidly coupled. The stators can rotate independently of one another. The following adjustment possibilities therefore occur:

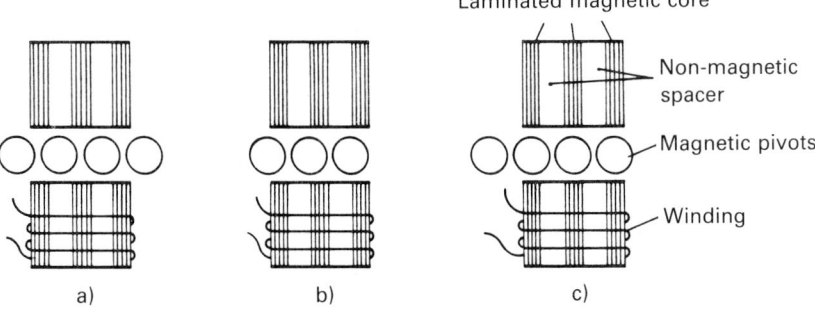

Figure 3.163  Position of the Accupin measuring head. (a) Minimum inductance. (b) Medium inductance. (c) Maximum inductance.

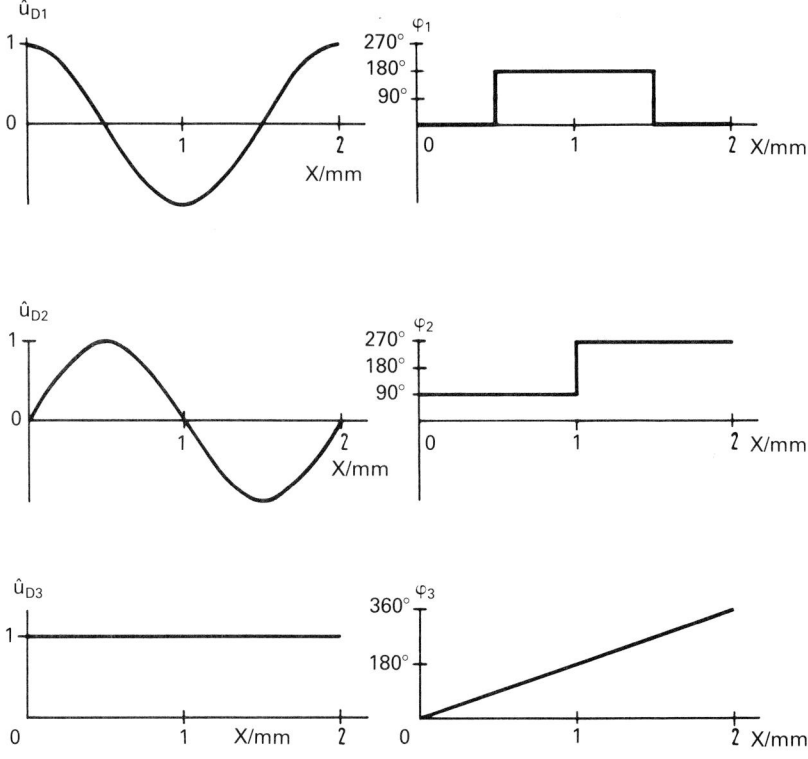

Figure 3.164   Amplitude and phase response of bridge voltages within a measuring cycle.

- adjustment of 90° phase shift between the two bridge feed voltages with simultaneous compensation of mechanical production tolerances in the measurement head through rotation of one stator
- adjustment of the electrical zero position of the system relative to mechanical zero setting through similar rotation of both rotors.

The complete system for position measurement by the phase-control process is shown in Figure 3.165. The output signal $U_A$ is read off via a null-seeking detector, which produces a square-wave signal, the phase-shift of which is measured. The accuracy of the measurement depends in the first place on three factors: accuracy of scale and measurement head, adjustment accuracy and accuracy of the analogue-to-digital converter. In the linear Accupin the maximum error is $\pm 5\ \mu$m. The circular Accupin with a rotor of 500 pins, arranged on a circle of 318 mm diameter, attains a maximum measurement error of ±3 arc seconds. Adjustment of the scale, which is made up of individual pieces of 6 or 10 inch in length, and adjustment of the measurement head have requirements similar to those of the Inductosyn. So, for example, the mounting surface of the scale must be aligned within 12.7 $\mu$m of the mounting surface of the measurement head. The tolerance for mounting the measurement head perpendicular to the surface of the scale is ±3 arc minutes. Similar requirements occur for the circular Accupin. For the permissible feed voltage, a range of 0–26 V$_{eff}$ applies. With a feed of 12 V$_{eff}$ the output voltage has a peak value of 0.3–0.5 V, and at the null-point a value of approximately 0.005 V. The permissible frequency range of 200–1500 Hz.

## 3.15   Static scanning of magnetized scales

A linear measurement system [148;149] with a scale which is magnetized like a magnetic tape can attain a sensitivity of 1 $\mu$m. The system consists of the scale, which is magnetized sinusoidally with a period of 0.4 mm, a magnetic head for readout and evaluation electronics. A notable property of the system is the fact that the scale can also be read when it is not in motion, Figure 3.166. To suppress the influence of magnetic interference fields, such as the Earth's magnetic field or magnetic fields with periodic variations other than those of the scale, several magnetic heads can be arranged at displacements of half a period relative to one another and electrically connected in series, Figure 3.167. The construction of the magnetic head is as shown in Figure 3.168. The readout has a non-linear transformer which is fed with a constant current $i_1$ with a frequency of 25 kHz (for binary systems) or 5 kHz (for decimal systems):

$$i_1 = \hat{i}_1 \sin\omega t$$

Depending on the magnetic field strength of the scale, the core of the transformer is more or less saturated, so that the output signal is modulated by the scale. The output signal contains a strong second harmonic which is limited by the hysteresis loop, which can be achieved either through small adjustments or through two parabolic loads (Rayleigh loop). The fundamental wave component arising from the excitation current is substantially compensated for by the arrangement of secondary coils around adjacent limbs of the magnetic core. An additional bandpass filter removes out the remaining

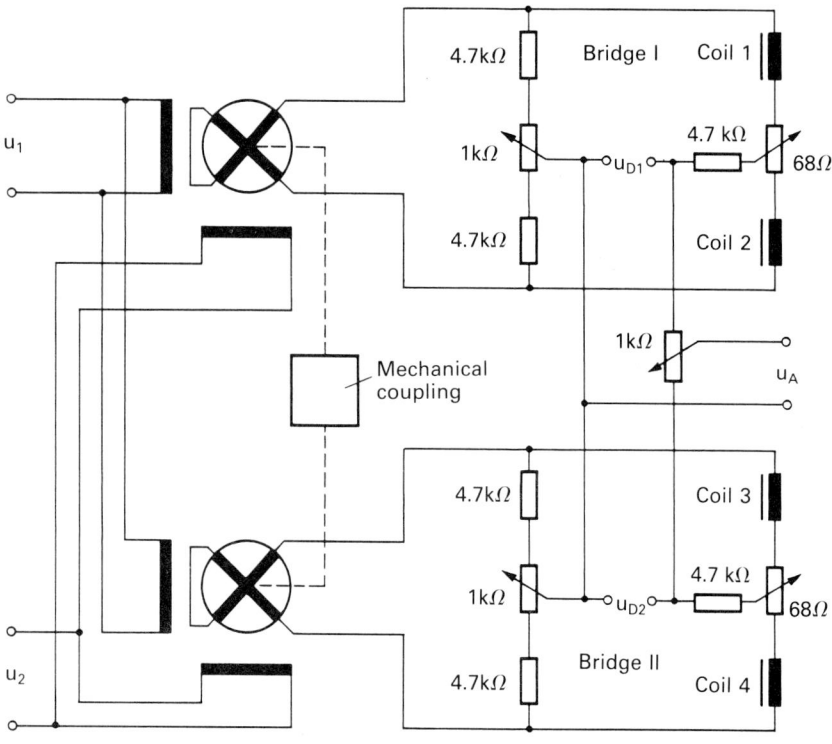

Figure 3.165   Basic circuit of Accupin measuring head.

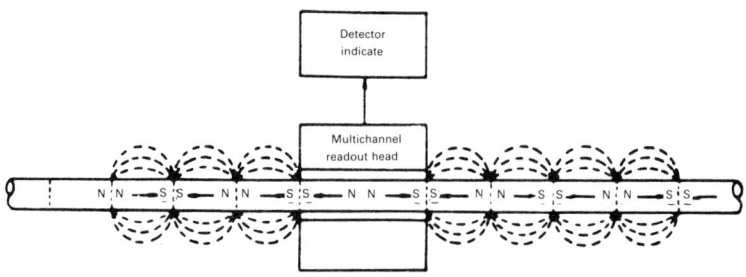

Figure 3.166   Principle of the magnetic scale measuring system.

components of the fundamental as well as the third and higher harmonics. The secondary voltage $u_{21}$ is expressed by the equation:

$$u_{21} = \hat{u}_2 \sin 2\omega t \cos \frac{2\pi x}{\lambda}$$

(123)

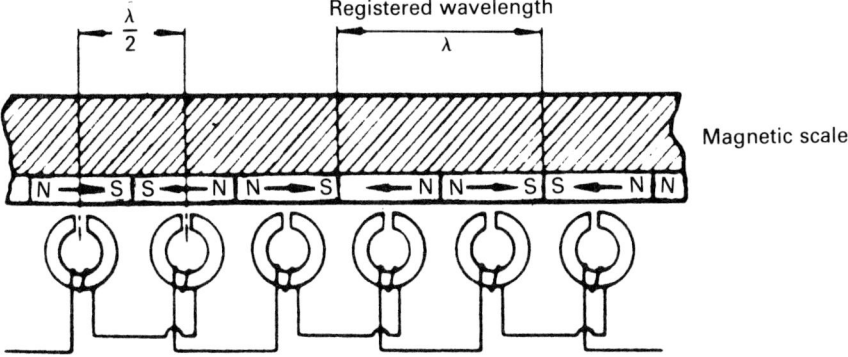

Figure 3.167    Principle of the multichannel head.

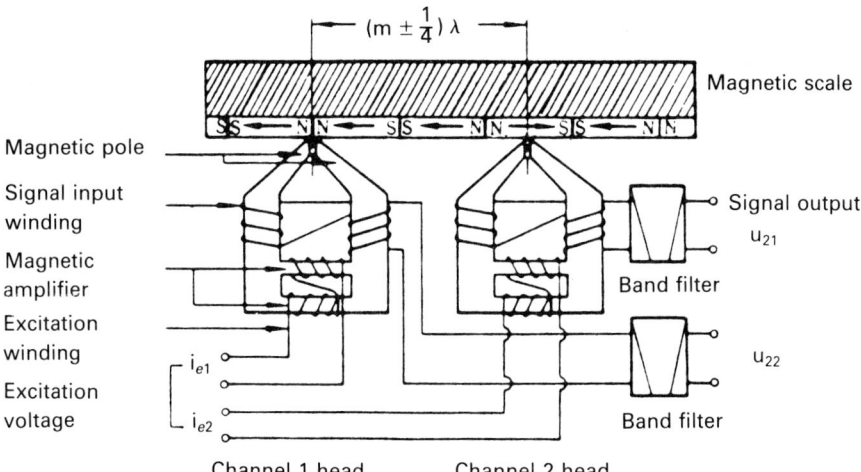

Figure 3.168    Schematic construction of the two magnetic heads used to generate out-of-phase signals.

The magnetic head 2 shown in Figure 3.168 is displaced from head 1 by $(m \pm \frac{1}{4})\lambda$. Here, $m$ is an arbitrary integer and $\lambda$ is the period of magnetization.

For the output signal we have:

$$u_{22} = \hat{u}_2 \sin 2\omega t \sin \frac{2\pi x}{\lambda} \tag{124}$$

This signal is phase-shifted by 90° by the time lag, and we then have:

$$u_{23} = \hat{u}_2 \sin(2\omega t - 90°) \sin \frac{2\pi x}{\lambda} \tag{125}$$

which corresponds to:

$$u_{23} = \hat{u}_2 \cos 2\omega t \sin \frac{2\pi x}{\lambda}$$

Then the sum $u_{21} + u_{23}$ is:

$$= \hat{u}_2 \left( \sin 2\omega t \cos \frac{2\pi x}{\lambda} + \cos 2\omega t \sin \frac{2\pi x}{\lambda} \right) \tag{126}$$

$$= \hat{u}^2 \sin \left( 2\omega t + \frac{2\pi x}{\lambda} \right) \tag{127}$$

by which we obtain, in place of an amplitude-modulated signal, a phase-modulated signal, the phase of which is a measure of the relative displacement of the readout head in relation to the scale. The phase displacement is digitally evaluated and gives the required sensitivity of $1\,\mu$m.

# 4

---

## *Intelligent sensors*

## 4.1   General properties of intelligent sensors

Intelligent sensors differ from conventional sensors in that they have
the ability to process the measurement signal within the sensor [150].
A conventional sensor for the measurement of angles is, for example,
a rotary potentiometer, which produces a voltage proportional to
the angular position of the disc under the no-load condition. If a
potentiometer is integrated with an amplifier, an analogue-to-digital
converter, a microcomputer and a power supply in a single unit, then
what we have is an intelligent sensor. The microcomputer with its
software enables the system errors of the potentiometer, such as non-
linearity and temperature-dependence, to be reduced.

For the measurement of temperature an additional temperature
sensor is required, and it is also possible with intelligent sensors to
cancel out error currents within the sensor itself (Figure 4.1). The

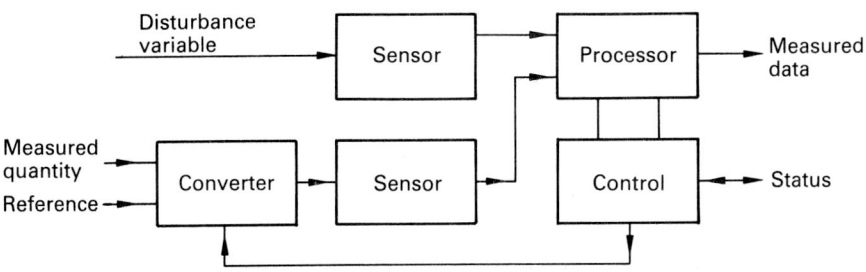

Figure 4.1   Structure of a measuring system with disturbance-variable
compensation and self-test routine.

result of the error corrector is an increase in the usable information output of the sensor.

With an analogue sensor, such as a laminar potentiometer, the sensitivity is theoretically infinitely high. With digital reprocessing, the analogue value must first be digitized, and then quantization is carried out. Here the maximum detectable number of steps $m$ is given by the magnitude of the static error $F$. For estimation of $m$ the following equation can be used:

$$m = 1 + \frac{1}{2F}$$

If, for example, the static error $F = 0.5\%$, then a maximum of 101 steps are detectable.

The information content is defined as $I = \log_2 m$ (bits).

As an example $I = \log_2 101 = 6.66$ bits. If in the intelligent sensor the static error of 0.5% is reduced by, for example 0.1%, then the maximum detectable number of steps will be increased by

$$m_1\ 1 + 1/(2 \times 0.001) = 1 + 500 = 501$$

The information content $I$ is therefore

$$I = \log_2 501 = 8.97 \text{ bits}$$

The maximum usable information content is thus approximately 2 bits or a factor of 4 greater.

A further interesting property of intelligent sensors is the ability to filter out a number of individual information packages. As an example, a potentiometer used as an angle sensor can have a particular angular value which represents the beginning of a dangerous condition. All values which lie in the safe condition must not be exceeded. As soon as the dangerous condition is reached, the sensor gives out an appropriate message. A summary of the numerous properties of intelligent sensors produces the following list.

- Digitization and coding of the measured value; with incremental sensors the numerical process affects the coding.
- Carrying out plausibility checks.
  Boundary condition and gradient checks.
  Measured values which, because of the fundamental properties of the process to be measured may not be exceeded, are known and produce error messages. By using complex models the system can generate additional warning and simulation signals for self tests.
  Warnings (diagnoses) can be sent to the operator as part of the error analysis (Figure 4.1) [151].

- Automatic correction or compensation of system errors of the sensors such as non-linearity, drift, frequency-dependency, etc. This includes automatic self-calibration [152].
- Redundancy reduction and information compression.
  Data not of interest can be eliminated. The information is compressed by the extraction of parameters such as averaging processes, scatter determination, classification, abbreviation (e.g. floating-point expression), etc.
- Calculation of derived magnitude for information readout. In distance measurement systems for processing machinery, calculation operations such as zero-point corrections, tool radius corrections, millimetre–inch conversions, multiplication by $\pi$, etc., can be carried out.
- Storage measured values and their associated parameters. The possibility of storing large quantities of data improves operational usefulness by program-controlled user guidance. Numerous commands can be given with fewer operational elements.
- Operation of a data interface for data transmission.
  Formatting a data frame (address, measurement data, test and monitor data, sensor status).

Reception and evaluation of commands and data (parameters). A rapid twin-wire transmission in the form of a serial bus can connect distributed measurement and control systems to the central unit so that a simple error location is possible by breaking the circuit [152–155].

Digital data from transfer gives greater security than analogue in respect of interference. High interference currents, different earth potentials and earth loops as well as long leads are disadvantageous in both cases. With digital systems, however, there is the possibility of error recognition and consequent correction, which is not available with analogue signals.

- Self-monitoring can take place while test programs are running.
- Individual control functions can be transmitted.
  Calculation of deviations by comparison of theoretical with actual values. Issue of control commands, e.g. from signals to reduce drive speed when near theoretical value (pre-switching point).

The transformation of basic sensors into intelligent sensors is shown in Figure 4.2. Figure 4.2a shows the simplest form of a measurement

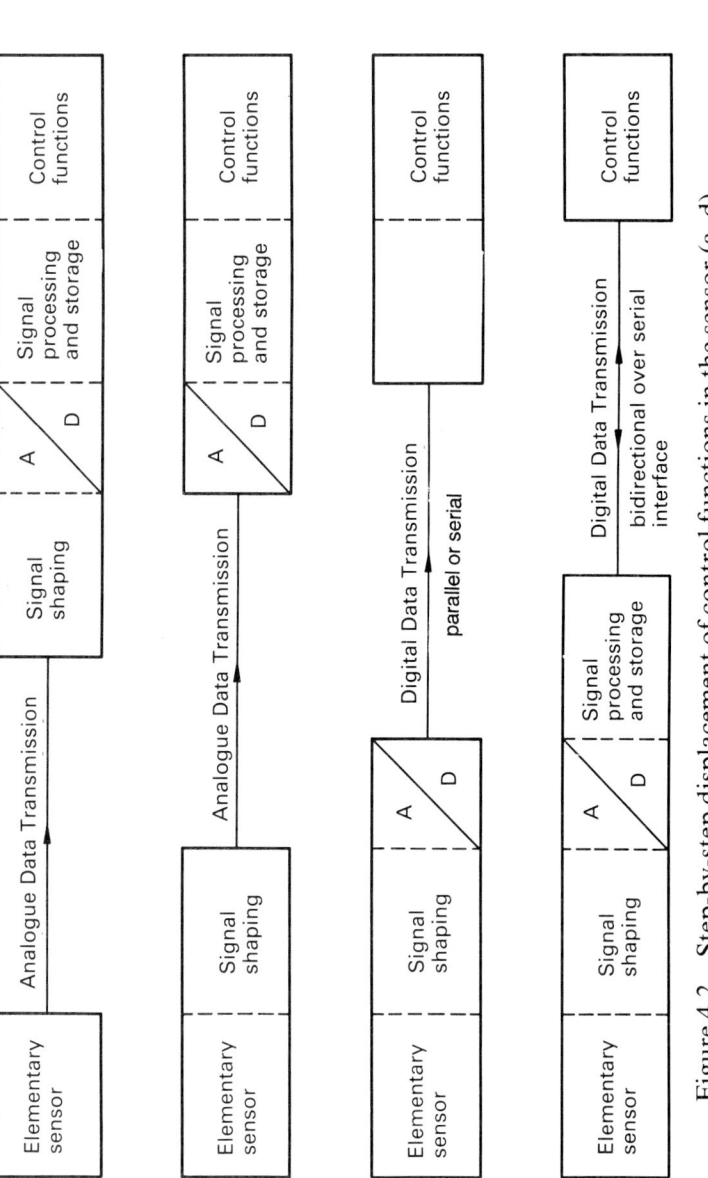

Figure 4.2 Step-by-step displacement of control functions in the sensor (a–d).

chain: an analogue sensor generates a measured value in, for example, the form of a current, which is first amplified in the control process and then digitized and processed in the digital computer. Sometimes analogue signals have to be amplified directly in the sensor before transmission (Figure 4.2). This applies in, for instance, capacitive position sensors. Because of the high internal resistance of the signal source, interference with the signal under direct transmission is almost unavoidable.

The following examples refer to sensors for both line and angle. Angles and distances are important physical dimensions for the control of motion of machines.

Figure 4.2c shows a sensor where the analogue-to-digital conversion is carried out within the sensor itself, as is the case with incremental angle step encoders and angle encoders. For all types of sensor, either parallel or series digital transfer of measured values is carried out. The step to an intelligent sensor is then shown in Figure 4.2d, since here the signal processing has already been carried out within the sensor.

## 4.2   Interconnection of intelligent sensors

Several physical and logical interfaces are possible for interconnecting sensors, and the following requirements are the most significant.

- The interface must conform to standard so far as possible in terms of both hardware and software, so that the sensor can work with distant stations or be used as a universal measured value source.
- Cabling costs must be minimal: this applies to series connections; for both buses and rings this condition is already fulfilled.
- Supply with a single voltage.
- High transmission rate and a minimum transmission range of 100 m.
- Economic interface components.
- Low development costs and simple checkability.
- Simple connection or disconnection of users.
- Defined handling of exceptional situations (accidents)

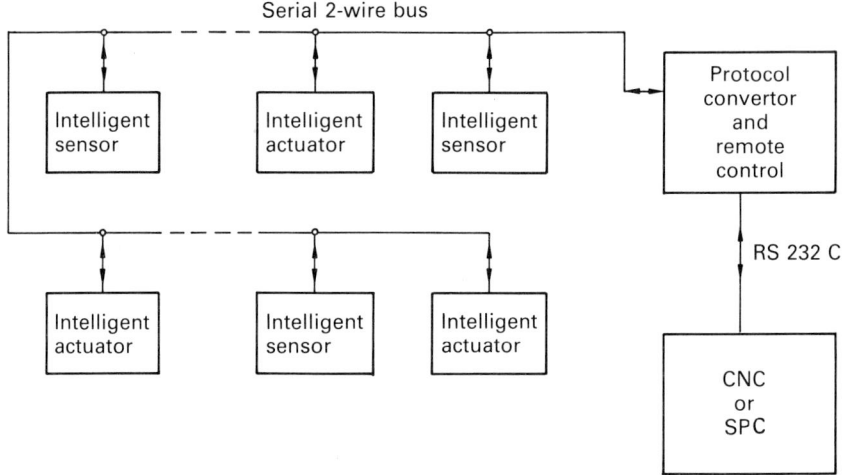

Figure 4.3    Several intelligent sensors and actuators coupled to a controller via an RS 485 data bus.

### 4.2.1   Series data transfer via a bus

The advantage of a bus (Figure 4.3) lies in the simple connection of new users and in the reduced cabling costs in comparison with conventional connections of users via a radial network [166].

Depending on its design, a bus can be suitable for either series or parallel data transmission. Well-known examples of parallel bus systems in computers and measurement and control systems are the Multibus, SMP or ECB systems. Because of their many parallel connections (typically 64 or more), they incur high technical costs when they are used over distances greater than about 50 cm, such as when several electronic units are interconnected in a single machine. For this reason, series bus systems have developed, which reduce cabling costs and allow transmission over distances of several kilometres. Each individual station receives all transmitted data and can send data in return. Because of the chronological transmission sequence of each bit, transmission speeds are, however, lower than with parallel bus systems.

### 4.2.1.1   General precautions against interference in series bus systems

Because it is necessary for individual users in a widely separated bus system to have their own decentralized auxiliary energy supplies,

Table 4.1  Examples of series bus systems with various transmission media

| Name | Transmission medium | No of subscribers |
|------|--------------------|--------------------|
| RS 232 C | Multicore cable | only 2 |
| RS 485 | Twin core cable | Up to 32 |
| Ethernet | Coaxial cable | Up to 1024 |
| Libsy | Optical fibre | Up to 512 |

differences in earth potential can cause interference. In order to overcome this problem, fixed specifications of the 'data transmission protocol' of the circuit must be observed in respect of the physical transmission medium (cabling, optical fibres, etc.), as well as the choice of data presentation.

In Table 4.1, typical users on series bus systems are listed, using different physical transmission media. Optical fibres are insensitive to differences in potential as well as to electromagnetic disturbances, but up to the present have only been used in exceptional cases in machinery because of their cost.

In the Ethernet system, data is input to or output from the circuit via a transformer, and so is protected through electrical separation.

At best, a series bus fulfils the criteria that apply according to RS 485 for an intelligent sensor for the measurement of line and angle (Figure 4.3). The EIA (Electrical Industries Association) has, in RS 485 standardized the electrical properties of receivers and senders in multipoint data transmission systems [156]. This standard is likely to be included as V.12 of DIN 66258 and DIN 66259 [157].

The most important properties of this bus are:

- only a single supply voltage (+5 V),
- transmits symmetrical difference signals and thus has high immunity to interference,
- high common mode rejection,
- simple interface components (SN 7516, SN 75177 or DS 3695, DS 3696); these consist partly of thermal protection equipment for short circuits in the driver output, and provide the possibility of automatic error localization through error indication signals,
- 32 senders, 32 receivers
- maximum transmission range 1200 m at 100 kBd,
- maximum data rate 10 MBd at 10 m range,
- semi or full duplex transmission (asynchronous),
- twisted cable suffices as a data transmission medium, although additional protection increases immunity to interference.

One disadvantage is the relatively small number of possible users. It should, however, be noted that a larger number is possible if only reliability is considered. However, with a greater number of users, the response time can become too great, since the concept of this local network does not provide for a user being active as such, e.g. CSMA/CD processes, but only for a slave user responding to a command from a master.

The master sends a block of variable length, but of at least 4 bytes, to a given user. Only this user must respond and send a block of variable length back to the master, see Figure 4.4. With electrically coupled systems, such as the RS 485 bus (Figure 4.5), the driver component must be in common with an earth lead of sufficient cross-section ($\geqslant 2.5\,\mathrm{mm^2}$) to equalize earth potential differences, so as to avoid exceeding the common-mode tolerance range.

The effect of electromagnetic interference on bus connections, such as arises in, for example, switching processes for contactors, etc., is countered in the relatively slow and short connections of the RS 232 C by a large voltage increase, the rate of increase of which is limited, and a safety zone between the dual conditions [160]. The immunity to interference with the RS 485 bus is at least an order of magnitude better [158;161;162].

The signal diagram of the RS 485 circuit results in a difference

| Address 5 bit<br>data length 3 bits | Data<br>1 – 8 bytes | Block protection<br>CRC 16 |
|---|---|---|

Figure 4.4 Block diagram of an RS 485 data bus.

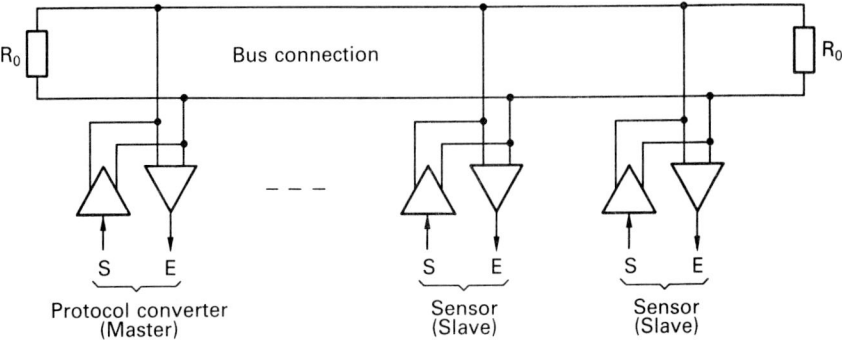

Figure 4.5 Physical design of RS 485 bus interface for multi-endpoint connections. S, Transmit; E, receive; $R_0$, terminal resistance.

signal. Interference pulses affect both circuits with the same polarity and are suppressed by the difference expression. For especially high requirements, the architecture of the bus system can be altered by converting to a ring or double-ring bus or by using a more expensive transfer protocol.

To the actual useful information coming from the address of a given user, which contains a message or measured value, a further two bytes of test information can be appended, formed according to the CRC (cyclic redundancy check) process (Figure 4.4). This process allows error detection with an accuracy greater than 99.998%. It is applied in various areas of technology, such as transmission of digital messages, testing electronic component groups (signature analysis) or in process control [164;165].

For each sending or receiving process, the test value is calculated from the useful information and attached to the information or compared with the received test value. Deviations between the test values indicate transmission errors, assuming that test value creation is itself error-free. From the information in the test value it is possible to apply a correction to the useful information transmitted; it is, however, usually quicker to obtain the information anew from the sender. A multiple CRC error with the same information indicates a cause other than random electromagnetic effects, such as, for example, hardware failure. Testing for multiple CRC errors must therefore take place in the receiving station in, for example, the control unit.

### 4.2.2 Bus protocol

Data exchange between stations must be clearly controlled and co-ordinated. This means establishing the length and format of the information to be transmitted through the bus circuit. Thus the word length, the sequence of data words and their significance, as well as their compilation, is established in a 'telegram'.

With the RSA 485 bus there is a data word format of 11 bits (1 start, 1 address, 8 data and 1 stop bit). This format is supported by new sender–receiver components, especially processors of the 8051 family, and allows a simple protocol with severe stations on the same bus (interprocessor communication). The address bit indicates whether an address or data value follows. All bus users check this but first with each telegram. If it signals an address, the address word is read and compared with its own 'house number'. If it is the same the data is intended for this user, and all other stations ignore all further data words until a new address bit indicates another address.

An additional speciality is a further bit in the address word, which indicates a general call to all system users. If this bit is read, the address information can be ignored. The message following represents a general call intended for all users of the bus system, e.g. switching all sensors in base condition. The next word transmits a command or a message, and the two bytes of the test word come after that. With the transmission of numerical values, the three numerical bytes precede the two test words.

## 4.3   An intelligent sensor for angle and distance measurement

Figure 4.6 shows the block diagram for the switching of an intelligent sensor. The capture and digitization of angle or distance information is achieved via a photoelectric angle step sensor, the pulse disc of which is scanned by a phototransistor. The light sources are photodiodes (LED). The output is a pair of signals, *A* and *B*, phase-shifted by 90° to obtain the number scan and directional information, and a zero pulse *N* emitted once per revolution. The two square-wave signals allow the resolution to be doubled or quadrupled very simply. Counting and processing the distance increments takes place in a 32-bit counter with direction sensor, separate from the microprocessor. The microcomputer itself cannot undertake this task for reasons of speed.

The microcomputer controls the progress of the operation, takes over supervisory data, such as a plausibility test of measured values relative to the zero signal, stores the counter condition in non-volatile memory (NVRAM) in case of voltage loss, and resets the counter on start-up or after an interruption to the value readout from the memory [167;168]. The NVRAM used has a capacity of 16 × 16 bits and serial interface. This gives additional write-protect security, in contrast to the parallel bus connection. There are also arithmetical functions, such as the four basic calculation operations used for correction (e.g. zero-displacement, miller radius correction: Figure 4.7), scale conversion (mm–inch and reverse) or the calculation of peripheral velocities from angular and radius values. Other important functions are the interrogation of three scans, two external contacts (key switch and end switch), the control of a 6-position

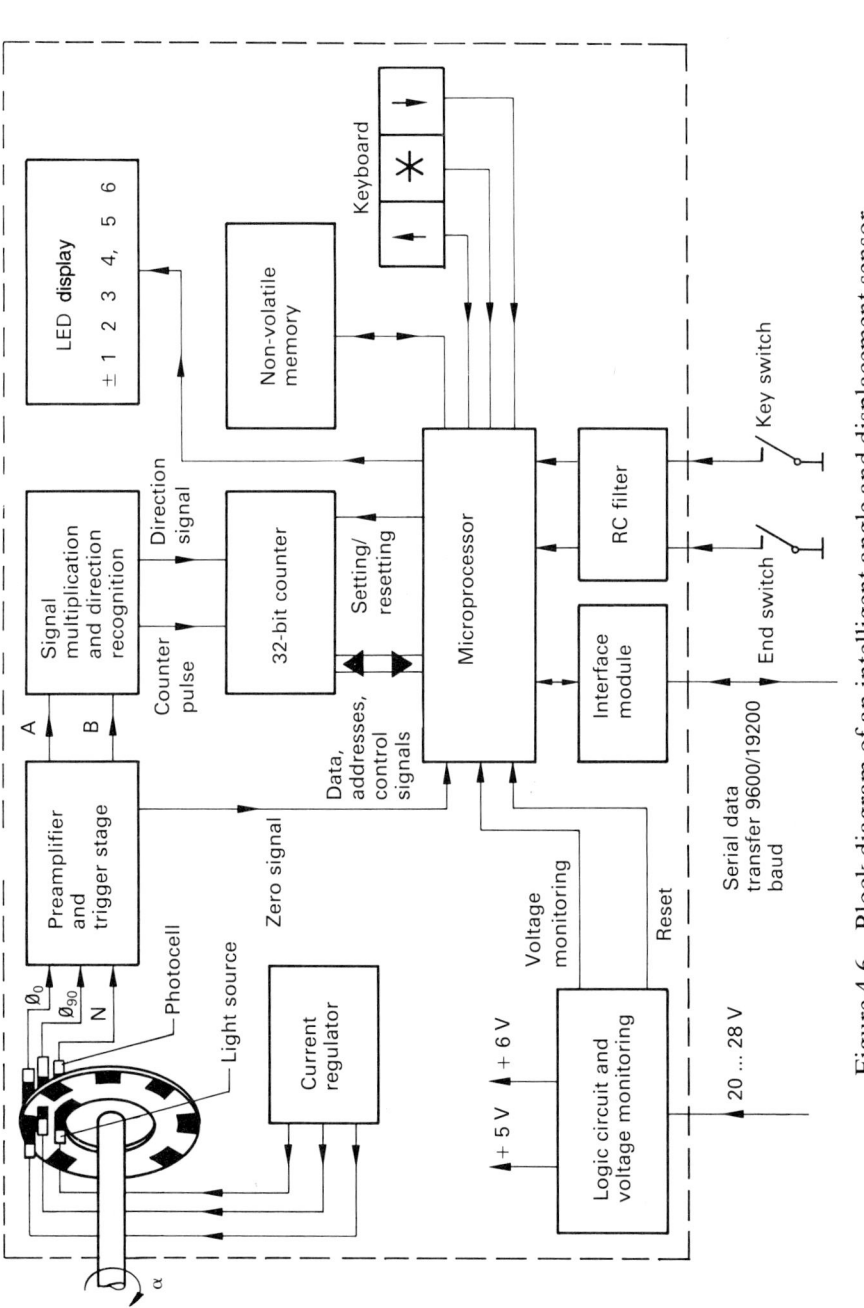

Figure 4.6   Block diagram of an intelligent angle and displacement sensor.

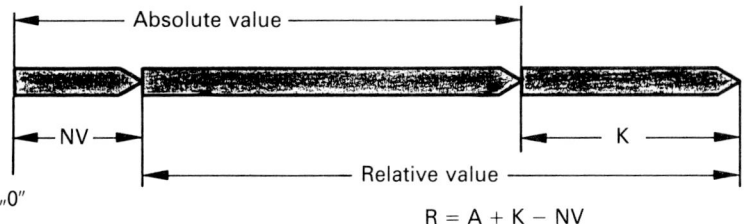

Figure 4.7   Example of calculation of relative value *R* from absolute value
*A*, zero displacement *NV* and tool compensation *K*.

display including algebraic sign and, not least, the transmission of
data via the serial interface.

A further important component is the built-in combinational
circuit (d.c. converter), which permits operation in the range 20–28 V
and thus removes the need for voltage regulation to a narrower
tolerance. At the same time, a signal is produced for low-voltage or
voltage loss (voltage check) and a signal for resetting the overall control system at start-up or after voltage loss [169–172].

# 5

---

## *Connection of measurement systems*

### 5.1 Mechanical links for converting longitudinal to angular motion

For length measurements rotary measurement systems are often used with intermediate switching and gearing or with ball-bearing spindles and nuts, which can offer cheaper and simpler solutions than direct linear measurement systems. Through the development of highly accurate linear-to-rotary converters, the disadvantages of indirect measurement in many applications have been overcome. The principle of the recirculating ball spindle or ball-bearing spindle has been known for a century, but was only brought into technical use between 1940 and 1950, when its wide potential was realized. Up till then conversion of rotary to linear motion was only possible via trapezoidal-thread shafts, the application of which was limited because of sliding friction. The recirculating ball shaft is a threaded shaft, in which, as the name implies, the balls run. The profile of the thread is either circular or oval. The connection to the nut, the profile of which is a mirror image of that of the shaft, occurs via the recirculating balls. Since the ball races are hardened and ground, friction is minimal, giving an efficiency of around 90% [173–176]. The design principle shown in Figure 5.1 occurs in two main forms for the recirculation of the balls as shown: the recirculating channel system, Figure 5.1a, which has the advantage of smaller size, since the nut is externally cylindrical and has no protruding parts, and the recirculating tube system, Figure 5.1b, in which the balls are introduced tangentially to the race and also leave it tangentially. The recirculating tube system has advantages in terms of wear of balls and races.

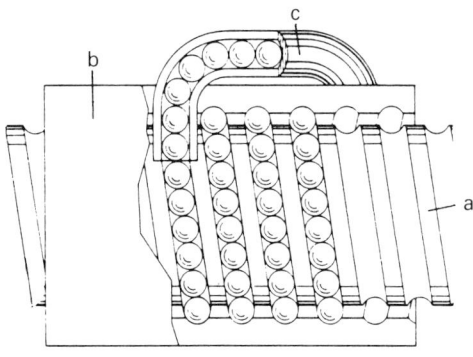

Recirculation system

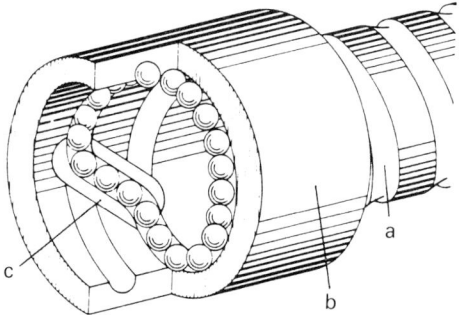

Figure 5.1    Ball recirculation system: recirculation unit system (above) and recirculation tube system. a, Spindle; b, nut; c, recirculation of balls. (Picture courtesy of Warner Electric Ltd.)

Figure 5.2 shows an opened-up recirculating ball nut in which the tangential entry and exit system can be clearly seen. Through the use of two counter-loaded nuts, the spindle drive can be made play-free. The spindle prestress in the region of the balls arising from this considerably reduces the elastic deformation under loading and therefore increases accuracy, Figure 5.3. The prestress usually forms about a third of the working loading since smaller values reduce stiffness and increase friction.

Developments in the range of recirculating ball drives extend to greater increase in load-bearing capacity, optimization of stiffness and reduction of friction [176].

The load-bearing capacity is improved by the use of balls designed for such higher loadings, an increased osculation between balls and races, as well as through the recirculation system and improvements

Figure 5.2   Ball screw with exploded view of nut. (Picture courtesy of A. Mannesmann Maschinenfabrik, Remscheid.)

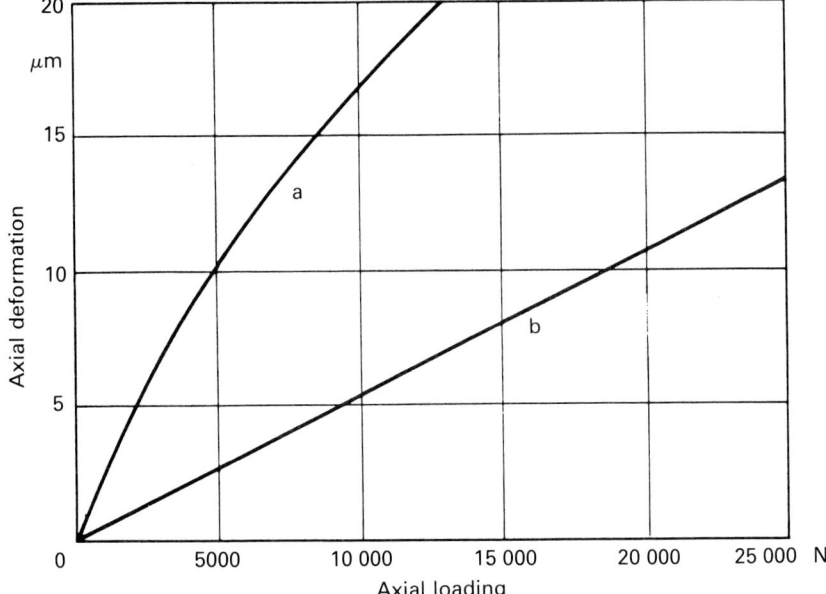

Figure 5.3   Effect of preload on elastic deformation of spindle. a, Not preloaded; b, Preloaded. (Picture courtesy of Rotax Precision Products.)

in the type and design of ball recirculation. There is also a role to be played by the connection between two counter-loaded nuts without an additional intermediate ring.

The reduction in rolling resistance and thus in friction means a reduction of heat developed, which again gives improvements in dimensional stability. It also reduces the stick–slip effect. Alternatively, a given permissible or preset nut friction increases the prestress and thus ensures increased stiffness.

Depending on taper, these shafts are available in five accuracy classes, based on a reference length of 300 mm for slope tolerance (DIN 69051 Part 3):

Accuracy Class 5
Accuracy Class 10
Accuracy Class 25
Accuracy Class 50
Accuracy Class 200

The number of the accuracy class is identical to the permitted taper in micrometers over a 300 mm thread length. Within each accuracy class, additional data determined from geometric tests can be given, defined as Categories A–E.

These values apply without load and at 20°C. In principle, the taper can be positive or negative and increases with spindle length. In practice, tapers at greater lengths are usually smaller than the theoretical value, and this gives rise to the possibility that summation errors can be reduced. Taper errors are then always negligible when the spindles undergo mechanical loading or are heated. A spindle of 50 mm diameter and 1500 mm length will deform by around $25 \mu m$ under a loading of 6800 N ($E = 2 \times 10^6$ N mm$^{-2}$. Apart from the influence on accuracy through thermal strain, recirculating ball spindles are suitable for a temperature range of −60 to +150°C. A second possibility for high-accuracy conversion from longitudinal to rotary movement is the rack-and-pinion [177;178]. Figure 5.4 shows components with a rack length of 250 mm and the appropriate pinion. By using several racks, any measurement length may be attained. For each element, a maximum summing tolerance of $5 \mu m$ (Grade 1) or $10 \mu m$ (Grade 2) is guaranteed. At the joint of two adjacent lengths is an equalizer for length variations, so that the errors do not cumulate. This level of accuracy is attained through high-accuracy grinding of the overall rack body, including the teeth. High surface quality and tooth flank hardness ensures long service life. The rack elements are usually fastened to the machine with clamps. In metric sizes, rack-and-pinion tooth pitch is usually 5 mm. The number of teeth on the pinion is selected so that one rotation corresponds to a measurement length of 100 mm. By dividing the pinion into two mutually stressed halves, it is ensured that the tooth flanks are always in contact with the teeth of the rack, thereby eliminating play. An example of the structures of a rack and a pinion is shown in Figure 5.5 [178]. By means of a Spieth spring collet, the pinion is mounted centrally on the axis of the sensor free of play. The sensitivity and range of the mechanical transmission ratio selected depends on the use of ball-bearing spindles or racks in the application of angle measurement systems. If this ratio is fixed by the design of the machine and there is no suitable sensor available, then the connection of the machine and data processing system must be via a gearbox, Figure 5.6. With indirect measurement through a measurement spindle and gearing, the relationship between the sensitivity with indirect measurement is as follows:

$$A_{ind} = A_{dir} \ddot{u} h \tag{127}$$

where $h$ is the spindle pitch, $\ddot{u}$ is the gear ratio $n_1/n_2$ and, for example, $A_{dir}$ 1/1000 r.p.m.

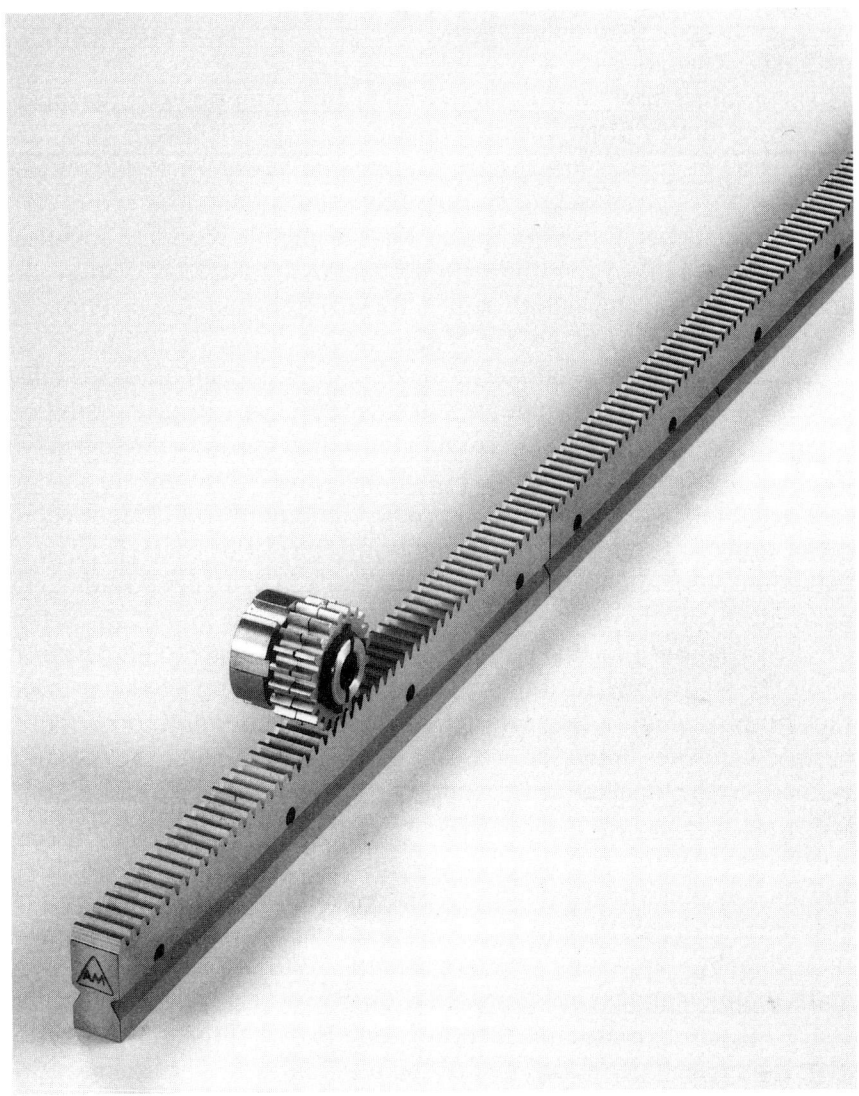

Figure 5.4   Length of measuring rack with preloaded pinion. (Picture courtesy of A. Mannesmann Maschinenfabrik, Remscheid.)

$$\ddot{u} = 2$$
$$h = 5\,\text{mm}\ U^{-1}$$
$$A_{\text{ind}} = (1/1000)\ U \times 2\ (5\,\text{mm}\ U^{-1}) = 0.01\,\text{mm}$$

The ratio of pinion to rack is similarly obtained. For the sensitivity in indirect measurement we have:

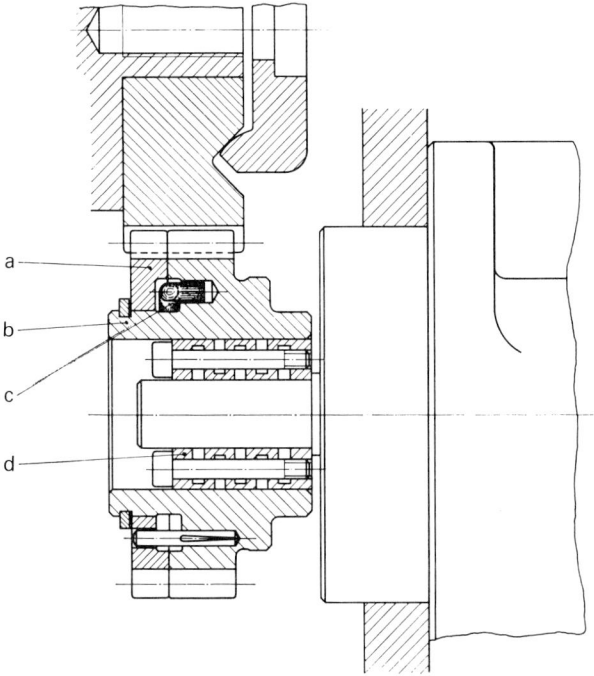

Figure 5.5 Example of measuring rack-and-pinion assembly. a, Preload pinion; b, measuring pinion; c, tension spring; d, Spieth bushing.

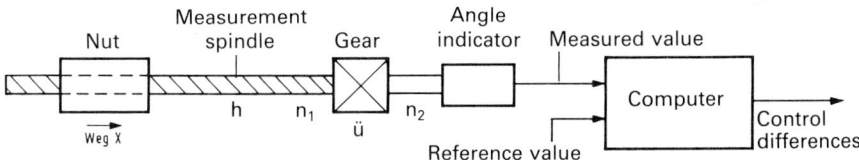

Figure 5.6 Using a gear mechanism to adapt a sensor to a machine and its arithmetic unit.

$$A_{\text{ind}} = A_{\text{dir}} \ddot{u} z t \qquad (128)$$

where $\ddot{u}$ is the gear ratio $n_1/n_2$, $z$ is the number of teeth on the pinion and $t$ is the pitch:

$$A_{\text{dir}} = (1/1000)\ U$$
$$\ddot{u} = 0.1$$
$$t = 5\,\text{mm}$$

$$z = 20U^{-1}$$
$$A_{ind} = (1/1000)\ U \times 0.1 \times 20U^{-1} \times 5\,mm = 0.01\,mm$$

In this case the modulus of the rack is $m = 1.5915$ (modulus = pitch/$2\pi$). The same sensitivity is obtained when the following combination is selected:

$$\ddot{u} = 1 : 15.708$$
$$t = 6.28$$
$$z = 25$$
$$m = 2$$

## 5.2  Mechanical attachments for angle measurement systems

The building of high-accuracy measurement systems into the object to be measured often requires attachments which ensure the transmission of measured data with the minimum possible error, as well as reduction of sources of error via production tolerances, including axial play, mismatching and angular deviation of the two shafts to be connected. In addition to the accuracy with which the angle is transmitted, the connection needs to have the maximum possible moment and the highest possible number of revolutions. For precision couplings used in fine working there are five essential design types:

- bellows couplings,
- helical couplings,
- Oldham couplings,
- claw couplings,
- membrane or disc spring couplings.

The bellows couplings, Figure 5.7, consists of two sleeves connected by a metallic bellows. It is cheap, maintenance-free and available at sensitivities to 0.01°. For high r.p.m. and at large construction tolerances between driven and sensor shafts, it is less suitable because of the high fatigue of the materials. Another disadvantage of these couplings is that they are relatively long and consist of a single integral unit.

The helical coupling as shown in Figure 5.8 resembles a tubular coupling, and also consists of a single inseparable unit. It is produced by cutting a helical groove in a cylindrical metal body. For this design the following properties occur: equalization of out-of-parallel in

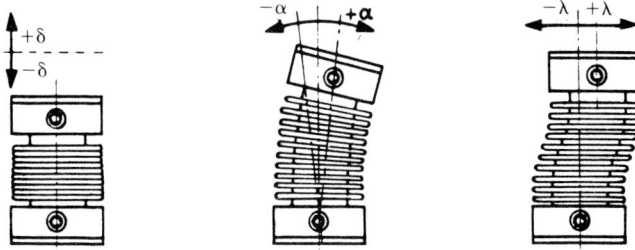

Figure 5.7 Bellows coupling to compensate for axial play, angle deviation and centre displacement of two shaft ends.

the connected shafts up to 3 mm or of angular error up to 30°; elimination of axial play and damping of torsional vibrations. Since the spring constant of the helical coupling is not changed by rotation, it gives equal bearing loadings. Speeds up to 25 000 r.p.m. are possible [179].

Simple construction, high torque capacity and the capability of accommodating large parallel shaft displacements at low revs (up to 250 r.p.m.) are the characteristics of the Oldham coupling [180;181]. All variants of this type of coupling consist of two discs, each with a transverse slot on its inner side, and a central torque transmitting disc

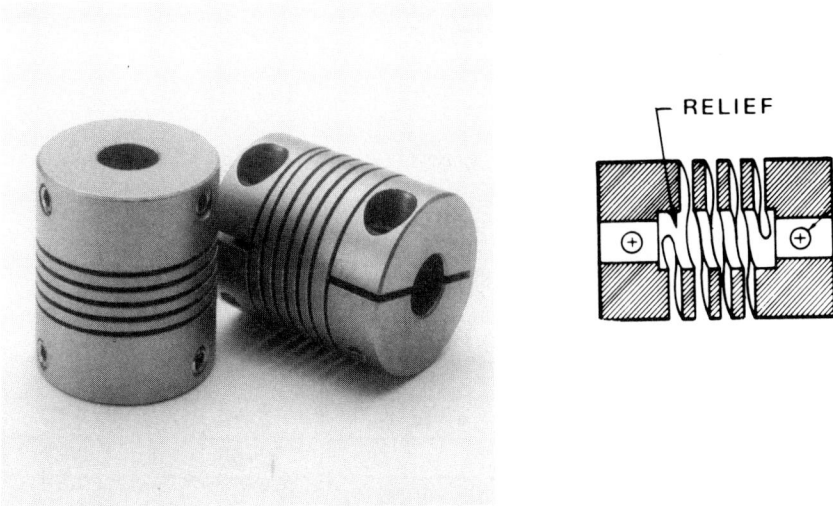

Figure 5.8 HELICAL coupling. (Picture courtesy of HELICAL Products Company Inc.)

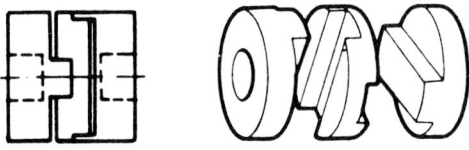

Figure 5.9    Principle of the Oldham coupling.

with matching slots on both sides, Figures 5.9 and 5.10. Angle matching is possible only up to 1°. A special variation of this type of
coupling allows electrical insulation between two machine components.

Membrane and spring couplings have a transmission ring in which
spring-loaded discs are arranged in the radial and axial directions,
which are especially rigid in the direction of rotation [182]. The connection consists of four points displaced by 90° in the form of a
Cardan joint on the transmission ring. Since this linkage is free from
play and friction, there is no wear. For equalizing the out-of-parallel
of the two shafts, two spring-discs connected by an intermediate
piece are provided. In the same way angular corrections are made by
a second Cardan joint (Figure 5.11). In a high-accuracy precision
coupling of this type [183] with, for example, maximum out-of-
parallel 0.1 mm, maximum angular error 0.07 and a 0.5 N cm torque,

Figure 5.10    Oldham coupling. (Picture courtesy of Tobias Baeuerle and
Söhne.)

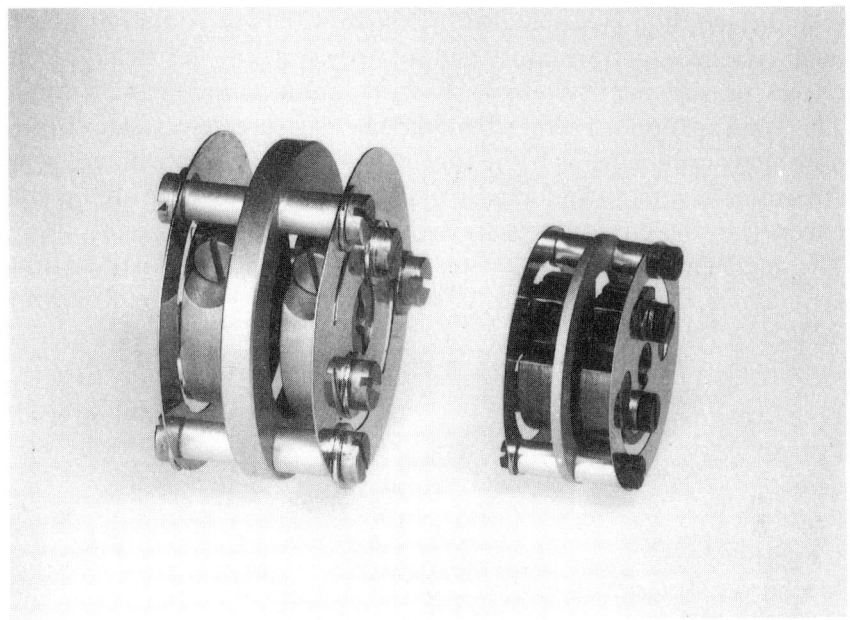

Figure 5.11   Spring lock couplings. (Picture courtesy of Tobias Baeuerle and Söhne.)

the following values are obtained: torsion error 0.5 seconds of arc, hysteresis error and kinematic angular transmission error of about 1 second of arc. The hysteresis error arises because the torsion in the coupling at maximum permissible torque does not return precisely to zero when load is removed. A considerable part of the kinematic error is due to the Cardan joint, and arises at angular differences between the two shafts. At constant r.p.m. of the driveshaft, the speed of the driven shaft varies slightly during each revolution, with a frequency of variation which is twice that of the frequency of the shaft. A special design of the coupling allows an adjustment of the angle of rotation with the coupling in place. After the coarse alignment during fitting, the zero position of the sensor can therefore be adjusted with the machine at its own zero position.

## 5.3   Electrical couplings for fine and coarse measurement systems

In a similar way to the transmission of signals from gear-coupled code discs in angular encoders (for a measurement range of several revolutions), it is possible also to couple widely separated measurement systems, e.g. translatory and rotatory systems, electrically, if the coarse measurement system works with double scanning and V-logic. This transmission is useful where the fine measurement system works absolutely cyclically, as is the case with a resolver or an Inductosyn. An example with an Inductosyn system superimposed on an angular encoder as the coarse measurement system should make this clear. The angle encoder is brought to the unloaded end of a ball-bearing spindle, which also serves to position it. It also gives the following data:

| | |
|---|---|
| Fine Measurement System: | Inductosyn with A–D Converter |
| Period: | 2 mm |
| Code: | BCD |
| Sensitivity: | $10^{-3}$ or $2\,\mu$m |
| Largest bit: | 1 mm |
| Mechanical Intermediate Component: | Ball-bearing Spindle |
| Diameter: | 50 mm |
| Length: | 1500 mm |
| Pitch: | 10 mm |
| Accuracy class: | 3 |
| Modulus of elasticity: | $22 \times 10^6\,$N mm$^{-2}$ |
| Loading: | 6800 N |
| Coarse Measurement System: | Angle Encoder |
| Code: | BCD |
| Sensitivity: | 5 positions per revolution |
| Measurement range: | 400 r.p.m. |
| Finest track on scanning line: | ±12.6° |

The tolerance range arising from the distance of ±12.6° between the two scanning elements of the first angular encoder track consists, over the shaft pitch, of a translatory movement around ±0.35 mm. The calculated ideal distance of the scanner is 1/20th of a revolution = 18°, or about 0.5 mm. From the actual ratios the following individual errors can be calculated:

| | | |
|---|---|---|
| Inductosyn: | Pitch error | $5\,\mu$m |
| | Adjustment error (6 linear) | $6\,\mu$m |
| A–D Converter: | Circularity error | $2\,\mu$m |
| Angle Encoder: | Scanning error | 1.5° / $29\,\mu$m |

Ball-Bearing
Spindle:

| | |
|---|---:|
| Pitch error | $150\,\mu$m |
| Axial play (nut and end-bearing) | $40\,\mu$m |
| Elastic deformation (6800 N) | $25\,\mu$m |
| Temperature error per 5°C | $85\,\mu$m |
| Total error | $342\,\mu$m |

Thus the maximum error which can occur, which can be positive or negative, lies within the permitted tolerance range and does not influence the measurement accuracy. This is the worst-case consideration, i.e. that all errors have the same sign and therefore can form a linear sum, but this method is preferred on the grounds of safety. The absolute numerical value can differ from case to case, so this should be seen as a rough approximation. A second example for the overlap of two systems is given in [184]. Here the increase in measurement range of a cyclically absolute system by mechanical switching is described. If the absolute measurement range of an angle encoder is increased from 10 to 20 m, then it is sufficient to mount two switches in a lead or lag position on the reading line with 10 m on the machine bed. These switches are operated by a notch on the moving machine part. A further extension to 40 m and so forth can be achieved in a similar way. The requirement is the use of two switches per track and combination of their signals via V-logic with the signals from the finer tracks or with those of the angle encoder.

# Bibliography

1. Das Internationale Einheitensystem und seine Anwendung. Arbeitsbl. No. 48. *Elektronik*, 1970, No. 7, pp 255–256.
2. DIN 1301, Einheiten, Einheitennamen, Einheitenzeichen, Nov. 1971.
3. Frühauf, G., *Praktikum Elektrische Messtechnik*. Braunschweig: Friedr. Viehweg & Sohn, 1970, pp. 4–5.
4. DIN 1319, Bl. 3, Grundbegriffe der Messtechnik, June 1981.
5. Zurmühl, R., *Praktische Mathematik für Ingenieure und Physiker*. 5th edn. Berlin, Heidelberg, New York: Springer Verlag, 1965, pp 287–292, 312–323.
6. Hoeschele Jr, D.F., *Analog to Digital and Digital to Analog Conversion Techniques*. New York, London, Sydney: Wiley & Sons, 1968, pp 15–24.
7. V D I 3254, Numerisch gesteuerte Werkzeugmaschinen. Genauigkeitsangaben Messen statistischer Kenngrössen, March 1971.
8. DIN 44 300, Informationsverarbeitung, Begriffe, March 1972.
9. Mütze, K., *ABC der Optik*, Hanau: Verlag Dausien, 1961, p 342.
10. Leinweber, P., *Taschenbuch der Längenmesstechnik*. Berlin, Göttingen, Heidelberg: Springer Verlag, 1954.
11. Woschni, E.G., *Informationstechnik*. Berlin: VEB Verlag Technik, 2nd edn, 1981, pp 234–235.
12. Schlichting, H., *Potentiometrische und induktive Weggeber*. Industrieelektrik und elek- der Interkama, 1974.
13. *Contraves-Rechenkondensatoren*, Bulletin 6561–583d. Zürich: Contraves AG.
14. Company catalogue: Absoluter Winkelaufnehmer DWG 82. Kirchzarten: Erwin Halstrup Multur GmbH, 1992.
15. Jones, B.E., *Messgeräte Messverfahren Messysteme, Part 1*. Munich, Vienna: R. Oldenbourg Verlag, 1980, pp 117–121.

16. Stöckl, M. and K.H. Winterling, *Elektrische Messtechnik*. Stuttgart: Verlag B.G. Teubner, 1978, pp 284–286.

17. Herceg, E.E., *Handbook of Measurement and Control*. Pennsauken, NY: Schaevitz Engineering, 1976.

18. Krause, D., Robuste Wegsensoren für extreme Belastungen. *Techn. Messen* **50** 1983, H. 10, pp 373–379.

19. Becker, K.P., Drehsinn erkannt. Vierfachauswertung für inkrementale Impulsgeber. *Elektronik* **8** 1984, pp 97–98.

20. Doktor, F. and J. Steinhauer, *Digitale Elektronik in der Messtechnik und Datenverarbeitung*, Vol. II. Philips Fachbücher, 1975, pp 96–100.

21. Voldan, W., *Using PAL (Programmable Array Logic) to get reliable digital data from SHAFT ENCODER CIRCUITS*. Company publication from Monolithic Memories.

22. Allen, T., Direction detector doubles as decoder. *EDN*, Sept. 17, 1987, p. 247.

23. Dart, A., Improved tachometer eliminates backlash. *EDN*, March 31, 1987, p 210.

24. Whittle, R.G. and M. Clarke, Photoelektrische Winkelcodierer— Trend zur Digitaltechnik und hochgenauen Auflösung. *Elektronik-J.*, Nov., 1970.

25. Elektronik-Bausteinsystem für das ZEISS PHOCOSYN-Verfahren. *Zeiss-Informationen* **80**, Special Supplement.

26. Tyler, J. Grundlagen der Fernmessung mit Pulskodemodulation. *Orbit* **6**, 1971, No. 4, pp 23–37.

27. Seidler, P., Nebenmaximumreduktion bei Korrelationsempfang binär phasencodierter Impulssignale. *Nachrichtentechn. Z.* **29**, 1976, H. 2, pp 154–159.

28. Brünnler, W., Winkelcodierer, Technik und Anwendung. *Siemens-Z.*, **44**, 1970, No. 4, pp 246–48.

29. Kerkow, W. and H. Böttcher, *Winkelcodierer mit Trägerfrequenzabtastung*. AEG Bulletin: Messen, Steuern, Regeln, Automation, 1964, pp 110–13.

30. Hoeschele, D.F., *Analog to Digital and Digital to Analog Conversion Techniques*. New York, London, Sydney: Wiley & Sons, 1968, pp 342–44.

31. Brünnler, W., G. Dorsch and W. Wagenberger, Winkelcodierer mit Hallgeneratorabtastung. *Simenes-Z.* **45**, 1971, No. 4, pp 259–260.

32. Kuhrt, E. and H. Lippmann, *Hallgeneratoren-Eigenschaften und Anwendungen*. Berlin, Heidelberg, Göttingen: Springer Verlag, 1968.

33. Burch, J.M., *The Metrological Applications of Diffraction Gratings. Progress in Optics, Vol. II*. Amsterdam: North-Holland Publishing Co., 1963.

34. Hock, F., Moirétechnik, Stand und Entwicklungstendenzen. *Maschinenmarkt* **46**, 1963, pp 23–33.

35. Pabst, W., Digitale Lagemessung an Arbeitsmaschinen. *Arch. f. techn. Messen*, 1963, pp R113–128.
36. Bonfig, K.W., Grundlagen und Eigenschaften der Lumineszenzdioden. ATM. *Arch. f. techn. Messen* **Z59–2**, 1975, pp 203–206.
37. Bergt, H.E., Optoelektronik—Blickpunkt der Halbleitertechnik. *Elektronikpraxis* No. 1/2, 1973, pp 7–11.
38. Fellner-Feldegg, H., Längenmessungen mit Lasern. *Umsch*, 1971, No. 12, p 433.
39. Hock, F., H. Kindl, W. Latzin and G. Sautter, Laser-Interferometer, ein Gerät zum Ausmessen und Justieren von Werkzeugmaschinen. *Siemens-Z*. **44**, 1970, No. 9, pp 549–55.
40. Dukes, J.N. and G.B. Gordon, A two-hundred-foot yardstick with graduations every microinch. *Hewlett-Packard Journal*, 1970, pp 2–8.
41. Rudé, A.F. and J.W. Kenneth, A new tool for old measurements—and new ones too. *Hewlett-Packard Journal*, Aug. 1970, p 9.
42. Metschl, E.C., Photoelektronische (optoelektronische) Bauelemente. *Elektronik-Industrie* **7/8**, 1972, pp 118–122.
43. Metz, S., Eigenschaften und Entwicklungstendenzen schneller Photodetektoren für die optische Nachrichtentechnik. *Nachrichtentechn. Z.* **29**, 1976, pp 127–133.
44. Scmidt, W. and O. Feustel, *Optoelektronik*. Würzburg: Vogel-Verlag, 1975, pp 53–105.
45. Company catalogue: NC-Längenmessysteme. Dr Johannes Heidenhain, Sept. 1990.
46. Sayce, L.A. and R.M. Pettigrew, Vorrichtung zum Messen der Verschiebung eines ersten Elements bezüglich eines zweiten. German Patent Application 2511350, 14.3.75.
47. Kordulla, H., D. Jansen and G. Schmid, Abtastkopf für Schrittgeber, insbesondere Winkelschrittgeber. German Patent Application DE 3444878 A1, 8.12.84.
48. Weyrauch, A., Zeiss Phocosin, eine neue lichtelektrische Digitalisierungseinrichtung für Längenmessungen mit $0.1\,\mu m$ Digitalschritten. *Zeiss-Information* **77**, 1971, pp 82–85.
49. *Linear-Messystem PE 2432*, Publication No. 7320.02.0115.13. Eindhoven: Philips.
50. Bremer, J.G., Linearmessystem für Werkzeugmaschinen. *ZWF* **75**, 1980, pp 351–355.
51. Spies, A., Längen in der Ultrapräzisionstechnik messen. *Feinwerktechnik & Messtechnik* **98**, 1990, 10.
52. Ernst, A., *Längenmessung im Sub-Mikrometer-Bereich*. Precision No. 1/1988. Zürich: Vogt-Schild AG.
53. Sesselmann, Th., Masstäbe für interferielles Abtasten ermöglichen Nanometer-Messchritte. *Werkstatt und Betrieb* **124**, 1191, 2, pp 111–114.

54. Hock, F. and K. Heitmann, *Fotoelektrischer Schrittgeber mit Fein-messeinrichtung.* DT-AS 1548 704, July 21, 1966.
55. Hock, F., Photoelektrische Messung der änderung von Längen- oder Winkelpositionen mit Hilfe von Beugungsgittern. Dissertation, Stuttgart University, 1976.
56. Götz, E., Direkt, digital und absolut arbeitende Lagemessysteme. *Steuerungstechn.*, 1968, No. 6, pp 224–29.
57. Klein, P.E., Winkelabtastung und -codierun. *Elektronik* **17**, 1968, No. 9, pp 283–86.
58. Rossi, B., Fehlererkennung und Fehlerkorrektur bei der Verarbeitung digitaler Information. *Elektroniker* **1**, 1970, pp 13–18.
59. Peterson, W.W., *Error-Correcting Codes.* New York, London: MIT Press and Wiley & Sons, 1961.
60. Steinbuch, K., *Taschenbuch für Nachrichtenverarbeitung.* 2nd edn. Berlin, Heidelberg, New York: Springer Verlag, 1967, pp 712–19.
61. Raible, B., Binärcodierte Masstäbe. *Elektronik* **16**, 1967, No. 3, pp 71/74.
62. Steinbuch, K., *Taschenbuch für Nachrichtenverarbeitung.* 2nd edn. Berlin, Heidelberg, New York: Springer Verlag, 1967, pp 56–61 and 66–67.
63. Janning, W., Codierung digitaler Informationen. *Automatik* **13**, 1968, No. 2, pp 50–56.
64. Wilhelm, J., Dreigitterschrittgeber, photoelektrische Aufnehmer zur Messung von Lageänderungen. Dissertation, University of Hanover, 1978.
65. *Digitiser Typenreihe 2800 Fotoelektrischer Winkelkodierer.* Company publication Dig. 2800 (April 1974) Fraba.
66. *Code pattern used in Rotax-Baldwin encoders.* Company publication of Rotax Ltd (1968).
67. Kliever, W.H., Measure position digitally. *Control Engineering* **3**, 1956, No. 11, pp 107–113.
68. Walcher, H., Die Codierung digitaler Lagemesseinrichtungen. *Arch. Techn. Messen* **427**, 1971, pp R89–96.
69. Wätzig, R., Schaltungsanordnung zur binären U-Ablesung eines dezimaltetradisch-codierten Lineals. German Patent Application 1258126, Feb. 8, 1964.
70. Götz, E. and W. Pabst, Einrichtung zur Stellungmessung an Arbeitsmaschinen mit einem dezimal-binär-codierten Code-Masstab. German Patent Application 1189626, Jan. 19, 1964.
71. Walcher, H., Ein neuartiges Abtastverfahren für Codemasstäbe. *Elektronik* **21**, 1972, No. 5, pp 151–54.
72. Walcher, H., Anordnung zur Abtastung eines codierten Masstabes, DT-OS 2041 832.
73. Walcher, H., H. Kreiling and B. Steffens, Der Winkelcodierer als hochgenaues, zuverlässiges und zugleich robustes Lagemessystem. *Feinwerktechn. Micronic* **76**, 1972, No. 4, pp 172–77.

74. Buck, S., Absolutwert—Winkelcodierer. German Patent Application DE 3322897 A1, June 25, 1983.
75. Miller, W., Photoelektrische Längen- und Winkelmessysteme. *Messtechnische Information* **3**, 1972, Jan., pp 6–11. Company publication of Dr Johannes Heidenhain, Traunreut.
76. Company brochure: Mechanische Positionsanzeige MPA 90–10/90. Kirchzarten: Erwin Halstrup Multur GmbH.
77. Grange, L., Laserentfernungsmessung. Lecture in training course B3, 06, Anwendung des Lasers in der Ballistik und Aerodynamik, produced by Carl-Cranz Gesellschaft e.V., Oct. 79.
78. Company publication: Linear-Messystem MK IV 9498.1822.119.18. Kassel: Philips GmbH.
79. Company catalogue: Transsonar Wegaufnehmer BTL No. 507D. Publication 9001. Neuhausen: Gebhard Balluff GmbH & Co.
80. Company publication: Ultraschall Wegmessysteme. Munich: Messring.
81. Company brochure: Model E201, Ultrasonic Ranging Module. Hingham, Mass., USA: Massa Products Corp.
82. Company brochure: Model E200, Ultrasonic Ranging Module. Hingham, Mass., USA: Massa Products Corp.
83. Balluff. Ultraschall für die ungestörte Wegmessung. *MSR Magazin* **1**, 1990, No. 1, pp 16–17.
84. Dornhagen, U. Intelligente Abstandssenroen. *Sensor Magazin* **1**, 1990, pp 13–18.
85. Krisch, L. Genau und berührungslos. Grundlagen und praktische Anwendung der Ultraschall-Distanzmessung. *Techno-tip* **15**, No. 5, 1985, pp 38–42.
86. Gebhardt, W., Bonitz, F. and H. Woll, Das Phased Array als neuer elektronisch steuerbarer Ultraschallwandler in der Werkstoffprüfung. *FhG Berichte* **3**, 1978, pp 60–65.
87. Nöll, H. Messwertverarbeitung in Ultraschall-Füllstands-Messgeräten. *Techn. Messen* **51**, 1984, No. 9, pp 313–317.
88. Michalski, B. and W. Berger, Füllstandsmessung mittels Laufzeit von Ultraschall. *Technisches Messen* **51**, 1984, No. 9, pp 306–312.
89. Benz, F., Elektronische Entfernungsmessung mittels Lichtwellen, Part I. *Archiv für techn. Messen*, Paper V 1122–11, June 1968, pp 113–118.
90. Company brochure: LEM 1/17 Laser Rangefinder. Munich: Siemens AG.
91. Herbaugh, R.E., Precision laser tracks aircraft time-space position.
92. Das 2-frequente Laser-Phasenmessverfahren. *Elektronik* **19**, p 84.
93. Entfernungsmesser. *Elektronikpraxis* No. 3, February 1989, pp 98–101.
94. Schriever, D.W., Lasermasse, *Industrie—Elektrik + Elektronik* **33**, 1988, No. 5, pp 32–3.
95. Hock, F. and K. Heinecke, Automatisches Vermessen und Protokollieren von Präzisionsmasstäben durch fotoelektrisches Interferometer

und fotoelektrisches Mikroskop. *Maschinenmarkt* **71**, 1965, No. 37, pp 27–37.

96. Hock, F., Grundlagen der Laser-Interferenzlängenmessung. Text of course at Technical Academies of Esslingen and Wuppertal, 1971.

97. Dokumentation Laserinterferometrie in der Längenmesstechnik, VDI Report No. 548, March 1985.

98. Tangungsband Laserinterferometrie in der Längenmesstechnik. Conference in the PTB, Braunschweig, March 1985.

99. Company publication: Laser Messystem HP 5528 A. 02–5952–7190 GE. Hewlett-Packard, Sept. 1982.

100. Technische Beschreibung Laser Interferometer 5525 A: Ein neuer Weg in die Präzisionslängenmesstechnik. Nieder-Eschbach: Hewlett-Packard.

101. Rost, H. Das Laser-Interferometer also Weg-Messystem an Werkzeugmaschines. *Ind.-Elektrik + Elektronik* **17**, 1972, No. 15/16, pp 385–88.

102. Präzisionsmessungen mit dem Laser-Interferometer. *Messen + Prüfen*, 1972, Jan., pp 39–40.

103. Höfler, H. and E. Bergmann, Laserinterferometer zu interferometrischen Längenmessung. German Patent Application DE 3706347 A1, Feb. 27, 1987.

104. Selbach, H. and Lewin, A., Laserinterferometer zur Positions- und Schwingungsmessung. *Feinwerktechnik und Messtechnik* **96**, 1988, 1/2, pp 33–36.

105. Meiser, H.-P., W. Luhs and G. Litfin, Laserinterferometer in der industriellen Messtechnik. *Feinwerktechnik und Messtechnik* **96**, 1988, 10, pp 421–424.

106. Burgwald, G.M. and W.P. Krüger, An instand-on laser for length measurement. *Hewlett-Packard Journal*, 1970, Aug., pp 14–16.

107. Aronowitz, F., *The Laser Gyro in Laser Applications. Vol. 1*. H. Ross, 1971, pp 133–200.

108. Baumann, R. Ringlaserkreisel. German Patent Application DE 3150160 A1, Dec. 18, 1981.

109. Weckenmann, A. and C. Linhart, Absolutgeber zur Steuerung von Handhabungsmaschinen. *Technisches Messen* **51**, 1984, Vol. 5, pp 165–170.

110. Klement, E. and G. Schiffner, Drehsensor in integrierter Optik aufgebaut. *VDI Nachrichten*, 1982, No. 5, p 11.

111. Schiffner, G., Lichtleitfaser-Rotationssensor auf der Grundlage des Sagnac-Effects. *Siemens Forsch- u. Entw. Ber.* **9**, 1980, No. 1.

112. Auch, W., *Entwicklingsstand und -potential des faseroptischen Rotationssensors*. Bulletin. Stuttgart: Standard Elektrik Lorenz AG.

113. v. Fabeck, W. *Kreiselgeräte*. Würzburg: Vogel Verlag, 1980, pp 397–400.

114. Meyer, H.U., Company publication: Das SYLVAC-Mess-System. Rennens, Switzerland: SYLVAC Métrologie.

115. Reichl, M., Vielpoldrehmelder, Winkelcodierer und Schrittschaltmotoren. *Siemens-Z.* **42**, 1968, No. 4, pp 256–58.

116. Arbinger, H., Betriebseigenschaften von Drehmeldern. *Feinwerktechn.* **65**, 1961, No. 5, pp 181–88.

117. Wildermuth, E. Der Resolver—ein moderner Analogierechenbaustein. *Feinwerktechn.* **63**, 1959, No. 9, pp 307–317, No. 10, pp 369–373.

118. Feller, R., Synchros. *Elektroniker*, 1967, pp 243–250.

119. Wolfgarten, W. Die Anwendung der induktiven Ortsmessysteme Resolver und Inductosyn. *Messen + Prüfen*, 1970, Jan., pp 27–31, 1970, Feb., pp 119–124.

120. Pfeiffer, T., Digitalisierung des analogen Linear-Inductosyns. *Elektronik-Anz.* **3**, 1971, No. 8/9, pp 177–180.

121. Wolfgarten, W and A. Fend, Positionsmessysteme. Switzerland. *Maschinenmarkt*, 1971, No. 39, pp 66–69.

122. Geyer, W. Dynamische und statische Eigenschaften von ausgewählten analogen Lagemessystemen bei numerischen Werkzeugmaschinensteuerungen. *Regelungstechn.* **13**, 1965, No. 3, pp 134–138.

123. Simon, W. *Die numerische Steuerung von Werkzeugmaschinen.* 2nd edn. Munich: Hanser Verlag, 1970, pp 104–128.

124. Schmid, H., *Electronical Analog/Digital Conversion.* Van Nostrand Reinhold Co., 1970.

125. Schmid, H., An Electronic Design practical guide for synchro-to-digital converters. *Electronic Design* **6**, 1970, March, pp 178–185, 1970, April, pp 50–58.

126. Status Report Digital SINE COSINE Generator (DSCG). New York: Inductosyn Corp., 1969.

127. Roberts, F.G., Servo-style circuit speeds synchro-to-digital conversion. *Control Engineering* **15**, 1968, No 1, pp 65–67.

128. Roberts, F.G., Multiplexing technique cuts most of digital-to-resolver/synchro conversion. *Electronics*, 1972, June 19, pp 94–97.

129. Rebel, H. and H. Schröter, Untersuchung eines vollektronischen Synchro-Digital-Umsetzers. Part 1. *Elektronik*, 1971, No. 10, pp 333–336.

130. Roberts, F.G., Synchro-to-digital converters: pick the one that fits the job. *Electronics*, 1970, March, pp 116–119.

131. Hyatt, G.P. Solid-state synchro-to-digital converter. *Computer Design*, 1968, March, pp 48–53.

132. Haxl, K., Verfahren zur Umsetzung von Resolverspannungen in Digitalwerte. *Elektronik*, 1970, No. 8, pp 261–266.

133. Glantschnig, F., Hochgenaue Messsysteme an numerisch gesteuerten Präzisionswerkzeugmaschinen. *Neue Technik* **A4**, 1966, pp 239–250.

134. Eberle, M. and G. Waibel, Ein analog-absolutes Lagemssystem für die numerische Steuerung von Werkzeugmaschinen. *Siemens-Z.*, 1964, No. 9, pp 669–672.

135. *Inductosyn, Principles and Applications.* Valhalla: Farrand Controls, Inc., Publication ER No. 312, Sup. 1.1.68.

136. *Inductosyn Electrical Characteristics.* Engineering Report No. 387, Valhalla, NY: Farrand Controls Inc., Jan. 1962.
137. *Inductosyn Accuracy.* Valhalla, NY: Farrand Controls Inc., ER No. 312, Sup. 1.1.68.
138. *Inductosyn Instructions, Linear Metal Inductosyn.* Valhalla, NY: Farrand Controls Inc., Publication Q.C.R. 114, July 1967.
139. *Rotary Inductosyn Mechanical and Electrical Alignment Procedure. Glass and Metal.* Publication 12.1.67.
140. Trene, O., Resolver/Digital-Umsetzung: Der Dreh mit der Winkelmessung. *Industrie-Elektrik + Elektronik* 30, 1985, pp 62–69.
141. Peters, E.-G., Winkeldigitalisierung mit Resolver/Digital-Wandler-Platine. *Elektronikinformationen*, 1985, No. 11, pp 71–72.
142. Fleming, T., Synchro/resolver converters bring low cost and small size to motion-control systems. *EDN*, 1986, Oct. 30, pp 61–68.
143. Schupak, L., Circuit provides 16-bit R/D conversion. *EDN*, 1987, Aug. 20, pp 252–254.
144. Jarosinski, T., Single-chip uP controls resolver. *EDN*, 1988, Aug. 4, pp 209–213.
145. *Inductosyn Precision Linear & Rotary Position Transducer.* TIB 810. Valhalla, NY: Inductosyn International Corporation, Dec., 1986.
146. Resolver: Funktionsprinzip und Anwendungen. *Elektronik Informatione*, 1985, No. 6, pp 94–98.
147. *Accupin, elektromagnetisches Präzisionswegmess-System für lineare oder rotatorische Wegmessung.* General Electric Co., GEK-4748 B (6).
148. Company publication: SONY Magnescale Digitale Positionsanzeige. 10076a.1082. Hommel Handel GmbH.
149. Company publication: SONY Magnescale Längenmess-System. 10129.0383. Hommel Handel GmbH.
150. Tränkler H.-R., Messtechnik und Messignalverarbeitung. *Technisches Messen* 53, 1986, No. 3, pp 120–121.
151. Breimesser, F., Zuverlässige Messeinrichtungen durch Eigentest und automatische Korrektur. *VDI-Berichte* 566, 1985, pp 353–364.
152. Birkle, M., Softwareunterstütztes Messen. *VDI-Berichte* 566, 1985, pp 51–62.
153. Schneider, F. and H.-R. Tränkler, Der Einfluss der Sensoren auf die Struktur mikrorechnerorientierter Mess- und Automatisierungssysteme. *Technisches Messen* 53, 1986, No. 2, pp 66–70.
154. Brignell, J., Will sensors outsmart the programmable controllers? *Electrical Review* 218, No. 1, pp 26–27.
155. Kuntz, W. and H. Walcher, Busfähiger intelligenter Sensor für Winkel und Wege. *Technisches Messen* 53, 1986, No. 6, pp 229–235.
156. Pippenger, D., ICs extend RS-422 to multistation application. *EDN*, 1985, March 21, pp 181–188.
157. Schumny H. and E. Seiler, Aspekte der Eichfähigkeit, der Normung und der Messwertübertragung elektronischer Sensoren mit Intelligenz. *Technisches Messen* 53, 1986, No. 2, pp 60–65.

158. Abendroth, H.-P., Busfähige Schnittstelle nach RS 485/422. *Elektronik* **12**, 1984, pp 97–98.
159. Kafka, G., Schnittstellen für die Datenübertragung. *Elektronik*, 1984, Issue 25, pp 76–80.
160. DIN 66 020.
161. DIN 66 259, Part 3.
162. Wilson R., Hochleistungsnetze für Peripheriegeräte. *Elektronik* **19**, 1986, pp 146–148.
163. Wagner, T.F.D., Libsy-Pernet—Ein lokales Netz für den Einsatz in der Industrie. *Der Elektroniker*, 1986, No. 3, pp 41–44.
164. Wagner, M. and D. Leisengang, Signturanalyse in der Datenverarbeitung—Anwendungen und Beispiele. *Elektronik* **21**, 1983, pp 67–72.
165. Frohwerk, R.A., Signature analysis: a new digital field service method. *Hewlett-Packard Journal*, 1977, May, pp 9–15.
166. Färber, G., *Bussysteme*. Munich: Oldenbourg Verlag, 1984.
167. Obermeier, E. and H. Reichl, Messwerterfassungssysteme und Sensorprinzipien. *Elektronik*, 1979, Issue 26, pp 23–29.
168. Preissinger, H., Messwerterfassung mit Mikroprozessoren. *Techn. Messen* **47**, 1980, No. 9, pp 307–312.
169. Walcher, H. and R. Bartosz, Intelligente Sensoren. Ein Weg zur Leistungssteigerung von Maschinensteuerungen. *Elektronik*, 1987, Issue 23, pp 115–128.
170. Kuntz, W. and H. Walcher, *Intelligente Messwerterfassung an Bearbeitungsmaschinen*. Freiburg: Haufe Verlag, Der Innovationsberater, 1986, pp 2/1951–2/1988.
171. Walcher, H. Der Einfluss intelligenter Sensoren auf die Steuerung von Be und Verarbeitungsmaschinen. Sensoren, Technologie und Anwendung. *VDI Berichte* **677**, 1988, pp 189–193.
172. Walcher, H. Dezentrale Messwertverarbeitung mittels intelligenter Winkel-und Wegsensoren. *Sensor 88, Nuremberg*. Wissenschaft, Vol B, pp 55–66.
173. *Rotax-Kugelrollspindeln. G.302*, 2nd edn. Hemel Hempstead: Rotax Ltd., 1970.
174. *Warner Electric-Kugelgewindespindeln—ein Programm der Präzision*. Wolfshaben: Warner Electric GmbH, BIIE-d-7007-MCE, 1971.
175. Company publication: AM-Kugelgewindetrieb. Remscheid: A. Mannesmann Maschinenfabrik, 1991.
176. Neuer Kugelgewindetrieb: Bei geringer Reibung hohe Tragfähigkeit und grosse Steifigkeit. *Maschine + Werkzeug* 11–12, 1984, pp 37–38.
177. Company publication: AM-Zahnstangen. Remscheid: A. Mannesmann Maschinenfabrik, 1991.
178. Company catalogue: KST-Präzision-Messzahnstangen. Siegen: Kabelschlepp Trading GmbH, 1972.
179. Company catalogue: HELICAL FLEXURES, one piece rotating shaft flexible couplings. Santa Maria, USA: HELICAL Products Company Inc., CAP 90-A.

180. Hahn, K. *Die Feinwerktechnik*, Giessen: Verlag Pfanneberg, 1953, pp 309–310.
181. Stübner, K.I. and W. Rüggen, *Kupplungen, Einsatz und Berechnung.* Munich: Hanser Verlag, 1961, p 52.
182. Company publication: Wellenkupplungen. St Georgen: Tobias Baeurle & Söhne GmbH & Co. KG, 1991.
183. Ernst, A., Präzisions-Membrankupplungen, Schnappkupplungen, Kupplungen mit eingebauter Drehwinkeljustierung. *Messtechn. Inform.* **4**, 1972, July. Company publication: Dr Johnannes Heidenhain, Traunreut.
184. Lott, H.G.E., E. Götz and P. Boese, Anordnung zur digitalen Lagemesung im natürlichen Binärsystem. German Patent Application 1127 097, Dec. 30, 1959.

# Index

Page references to explanatory charts, tables and illustrations are set in italics.